T0257319

Early Praise for *Portable Python Projects*

This book covers many aspects of computer technology with a wonderful building-block approach, teaching the reader through every project. From the simple first project to the more complex, the builder is skillfully taught about Raspberry Pis and their hardware, Python and its libraries, automation, and most importantly, designing something useful from all these parts. By the end of the book, the reader is equipped to continue exploring in many new directions. I love this book.

➤ **Michael Bengtson**
Engineering Manager, Sprocks

I would recommend this book to most of the experienced software developers I know. It is a perfect read for long evenings in autumn when you have time to learn something new. This book can help you to improve your home, and it can inspire tons of new ideas you'd never thought of before reading it.

➤ **Maik Schmidt**
Software Developer and Writer

Portable Python Projects takes a practical approach to learning Raspberry Pi Python through complete implementations and concise meaningful explanations. I learned a number of useful Python strategies and a great deal about the Raspberry Pi platform. Python and the Pi are a perfect combination because together they are a simple, effective, and powerful platform. I highly recommend *Portable Python Projects* to both beginners and long-time enthusiasts.

➤ **John Cairns**
Staff Engineer, Blocknative

With this book Mike Riley opens the door for anyone who is curious about what can be done with a microcontroller and a little Python code. Newbies will find great step-by-step instruction to complete some innovative projects, and veterans will find inspiration to take take things to a new level.

➤ **Sven Davies**
   IT Director, AC Americas

Mike's insightful approach to practical Pi projects will allow new as well as experienced people to quickly build innovative solutions.

➤ **Jim Szubryt**
   Senior Delivery Principal, Slalom

Portable Python Projects
Run Your Home on a Raspberry Pi

Mike Riley

The Pragmatic Bookshelf

Raleigh, North Carolina

Many of the designations used by manufacturers and sellers to distinguish their products are claimed as trademarks. Where those designations appear in this book, and The Pragmatic Programmers, LLC was aware of a trademark claim, the designations have been printed in initial capital letters or in all capitals. The Pragmatic Starter Kit, The Pragmatic Programmer, Pragmatic Programming, Pragmatic Bookshelf, PragProg and the linking *g* device are trademarks of The Pragmatic Programmers, LLC.

Every precaution was taken in the preparation of this book. However, the publisher assumes no responsibility for errors or omissions, or for damages that may result from the use of information (including program listings) contained herein.

For our complete catalog of hands-on, practical, and Pragmatic content for software developers, please visit *https://pragprog.com*.

The team that produced this book includes:

CEO: Dave Rankin
COO: Janet Furlow
Managing Editor: Tammy Coron
Development Editor: Jacquelyn Carter
Copy Editor: L. Sakhi MacMillan
Indexing: Potomac Indexing, LLC
Layout: Gilson Graphics
Founders: Andy Hunt and Dave Thomas

For sales, volume licensing, and support, please contact *support@pragprog.com*.

For international rights, please contact *rights@pragprog.com*.

Copyright © 2022 The Pragmatic Programmers, LLC.

All rights reserved. No part of this publication may be reproduced, stored in a retrieval system, or transmitted, in any form, or by any means, electronic, mechanical, photocopying, recording, or otherwise, without the prior consent of the publisher.

ISBN-13: 978-1-68050-859-8
Book version: P1.0—February 2022

Contents

Part I — Setup

Part II — Projects

Acknowledgments

Writing a book is like having a child, in that only those who have one truly know what the experience is like. This is my fourth book published by Pragmatic Bookshelf, and like those before, it could not have been possible without the tremendous support and consistent encouragement and guidance of the editor and publishers. Jackie Carter has been the development editor on all my Pragmatic titles, and she once again has shown with this work why she is one of Pragmatic's best editors. Thank you Jackie for everything!

I would also like to thank those who provided invaluable feedback on my earlier drafts to bring more clarity and specificity to the projects. Mike Bengtson, John Cairns, Sven Davies, Jim Szubryt, John Winans, and fellow Pragmatic Bookshelf author Maik Schmidt were especially instrumental in guiding the book to the final form you are about to read. You guys are amazing!

Finally, it should go without saying that without Andy Hunt and Dave Thomas, the progenitors of the Pragmatic Bookshelf, this book would not exist. They are two of the smartest technologists and book authors in the tech industry and have built a trusted platform under the expert guidance of CEO Dave Rankin that is among the most respected tech book publishers in the industry. It is an honor to be associated with such a top-tier group of outstanding individuals.

Introduction

Welcome to the Pi. The journey ahead will guide you through the process of obtaining and configuring the right hardware and software to create unique and interesting applications.

A lot of books are written about the Pi, but nearly all of them focus on basic hardware and rudimentary electronics tasks that fail to showcase the power that the latest Pi hardware and add-ons have to offer. The base software we will configure on the Pi are the same applications used by world-class developers. And the projects we'll create offer a functional foundation that you can use to extend and customize for your own ideal applications.

The Raspberry Pi hardware has evolved considerably since its initial release in 2012. The fourth-generation design represents a major leap forward in bringing inexpensive yet powerful computers to the masses. The Pi 4 can support up to 8 GB RAM, which is twice as much as many desktops and laptops sold just a few years ago. More memory equates to more sophisticated applications running at the same time.

The syntax and structure of the Python language is easy to learn for beginners and yet is powerful enough for professional software developers to use every day. It is portable in that scripts written on one supported operating system can be reused with little or no changes on another OS.

The Raspberry Pi is also portable in that its size is smaller than a deck of playing cards, allowing it to be set up or easily relocated anywhere in your home or office. This is why the Pi is so popular among home-automation hobbyists. Thanks to the Pi's low-power CPU and robust compact design, it can be positioned behind a TV, near a washing machine and dryer, or even inside a light-switch compartment. As long as there's a power source and an Internet connection that the Pi can connect to, its location and use-case possibilities are endless.

About the Book

The first part of the book introduces the Raspberry Pi computer, the peripherals required to effectively operate it, and a number of nifty sensors that provide the Pi with extended data processing capabilities. We'll also install the operating system and other software needed to prepare the Pi for the projects in the book.

The second part of the book contains the projects. Each project is similarly presented, starting with a brief introduction describing the objective. The Setup section contains all the additional hardware and accessories required to construct the project as well as any additional software libraries and packages needed for that project. The next sections walk you through the assembly of the components and the Python scripts needed to make them do something interesting. Projects conclude with a Next Steps section encouraging you to customize and expand the project foundation into a solution that ideally suits your needs.

Online Resources

The apps and examples shown in this book can be found at the Pragmatic Programmers website for this book.[1] There you will also find the errata-submission form, where you can report problems with the text or make suggestions for future versions.

In the next chapter, we'll review the shopping list of recommended hardware used in the book's projects. In addition to price and features, we'll also discuss the rationale for using one hardware configuration over another. So get ready for an educational, fun, practical, and rewarding experience ahead!

Mike Riley
mike@mikeriley.com
January, 2022

---

1. http://pragprog.com/titles/mrpython/

Part I

Setup

Building any successful project requires preparation. In this part, we'll review the hardware and software needed to assemble our projects and then prepare them for use in Part II of the book.

CHAPTER 1

Assembling the Hardware

Making your own Pi projects is not only a rewarding, educational endeavor, it's also a lot of fun. It's enlightening and empowering to see your creations come to life after connecting inexpensive hardware and writing a few lines of Python code. In this book, you're going to create cool, practical projects, spanning water leak alerts to security camera captures, and home automation to making chatbots. Each of these projects builds on the next, so it's best to do them in order. After building the projects in this book, you'll have the ability to put your own amazing automation ideas into practice.

But before we can begin these Pi project adventures, you need to acquire the necessary hardware to construct them. This chapter will quickly review the components required to bring the Pi and the book's projects to life. Each item will be accompanied by the average price, the web links to purchase them, and a brief summary of what the hardware is and why you'll need it.

Building your own hardware and software provides considerably more flexibility, even though it takes more up-front planning and time commitment. In addition to a steeper learning curve, customized hardware is also an iterative process that often requires numerous prototypes before a satisfactory working solution is attained.

The projects in this book take a hybrid approach and achieve the best of both worlds. The hardware sensors we'll use are pre-built and don't require a degree in electronics to connect to and use with your Pi—no breadboard or soldering gun required. The software we'll write to run the projects can be highly tailored for your own particular needs.

For example, you could purchase a pre-built water leak detector with a built-in WiFi radio that uses a third-party service to alert you when a leak has occurred. But the cost of such a device could be more than the cost of the most expensive Pi hardware configuration and attached water leak sensor combined. Case in point—the Eve Water Guard[1] with the Eve Extend retails for $129.95. You'll build a nearly identical system for far less. And because the Pi is a multipurpose computer, it can simultaneously be a file server, a source code version repository, a multimedia player, or a desktop computer while checking for any water leaks.

Another benefit is, unlike third-party products that may send any variety of potentially sensitive network data to third-party servers that may or may not be trustworthy, you know exactly what data is being sent where because you wrote the code that instructs the Pi what to do with the data collected. It seems every week there are news reports of intentional backdoors, exploited software, abandoned services, and other security failures that compromise so-called third-party Internet of things (IoT) devices. Assembling your own Pi, running Python code you wrote, means your solution is no longer a mysterious black box. You no longer are forced to rely entirely upon the product's vendor for updating and patching against unexpected or even intentional vulnerabilities. With the Pi, you are in control.

Depending on your level of tech-savviness, you may already own several of these products. Also, keep in mind that you don't need to purchase every accessory listed in this chapter. You can pick and choose what hardware and accessories you want, depending on the projects you wish to pursue. Here are the hardware items you'll need to build the variety of projects in this book.

Raspberry Pi

Many of the projects in this book will work with models as old as the Raspberry Pi 2 or even the miniature Pi Zero. But the best Pi available at the time of this book's publication is the Raspberry Pi 4, and that's the model that will work for all the projects. It's also the model shown in the photos shown throughout the book, such as in the photo on page 5.

The Pi 4 model is available in four different memory configurations: 1, 2, 4 and 8 GB. If you plan on running only one project, or two projects simultaneously, the 1 GB ($30) or 2 GB models ($35) will suffice. The 4 GB version ($55) is the best compromise for price-conscious consumers looking for the best investment. But if you want to run several projects simultaneously on the Pi with room to spare for future projects of your own, the 8 GB model ($75) is recommended and the one I prefer for all my Pi-related projects. You

---

1. https://www.evehome.com/en/eve-water-guard

can purchase any of the Pi models from various Raspberry Pi Foundation–approved retailers. Visit the official Raspberry Pi Store[2] to select your Pi 4 model of choice from the list of retailers in your country.

microSD Card

The Pi can boot from microSD exFAT-formatted cards with storage up to 2 TB, but unless you plan on using your Pi for network-attached storage (NAS), I've found little need to expand beyond a 32 GB microSD card. That said, microSD cards are so inexpensive, with 128 GB cards costing below thirty dollars, that having extra space for media files, code repositories, and images can be useful. For example, the photo on page 6 shows a 32 GB microSD card ready to be inserted into the bottom of a Raspberry Pi 4 Model B.

Several retailers sell the Pi as a kit that includes a microSD card (usually 32 GB), power supply, micro-HDMI to HDMI converter cable, case enclosure, heat syncs and/or mini fans to dissipate heat generated by the Pi's Broadcom graphics chip, and ARM-based CPU. These kits are useful if you don't want to purchase these necessary components separately. I purchased my Pi 4

---

2.   https://www.raspberrypi.org/products/raspberry-pi-4-model-b/

8GB in a kit that included a power supply, micro-HDMI adapter, and micro-USB to USB-C adapter. The package only cost ten dollars above the price of the base 8 GB Pi model. Had I opted to buy these accessories separately, I would have spent more on just the Pi's power supply alone.

USB-C Power Supply

The Pi 4 Model B requires a 5.1-volt, 3-amp, 15.3-watt USB-C power supply to adequately power the Pi. The Pi 4 does not include a power supply unless it was purchased as part of a bundled kit. The first photo on page 7 is of a Raspberry Pi 4–compatible USB-C power supply obtained from a third-party Pi vendor.

Power plugs come in four variants, depending on whether you're in the United States, United Kingdom, Europe, or Asia. Visit the Raspberry Pi Store[3] to choose the type and the retailer to purchase it from.

Micro-HDMI Adapter

Unlike previous versions, the Pi 4 does not have a full-sized HDMI port. Instead, it offers two much smaller micro-HDMI ports to connect the Pi's audiovisual output to your computer monitor or TV. It's a testament to the power of the Pi 4 hardware that the device is capable of outputting video simultaneously on two 4K displays. If you're fortunate enough to have two

---

3.   https://www.raspberrypi.org/products/type-c-power-supply/

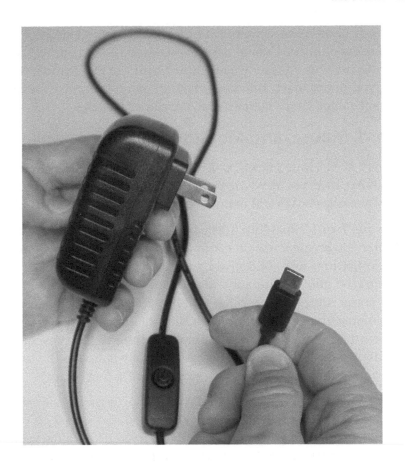

4K displays to connect to the Pi, you'll also need two micro-HDMI to HDMI adapters along with standard HDMI cables. Check out the a photo of a micro-HDMI to HDMI port adapter.

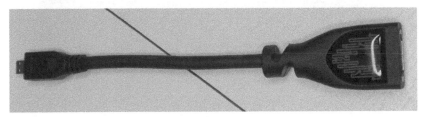

Just like the other Pi hardware, you can visit the micro-HDMI product page[4] at the Raspberry Pi Store to select the color and length of the cable and locate a retailer in your country to purchase it from.

---

4.   https://www.raspberrypi.org/products/micro-hdmi-to-standard-hdmi-a-cable/

Since I already had several spare HDMI cables lying around in my home, I opted instead to connect my Pi 4 using a micro-HDMI to HDMI adapter. As I previously mentioned, this adapter was included with the Pi 4 bundle that I purchased. If you prefer to get this kind of stand-alone adapter, you can buy it for about six dollars from retailers like Amazon or Newegg.

Keyboard, Mouse, and Monitor

Any standard USB computer keyboard and mouse can be used with the Pi. If you don't have an extra wired keyboard and mouse available, I recommend buying a wireless keyboard and mouse that use a dedicated 2.4 Hz USB dongle.

While you can pair a Bluetooth keyboard to the Pi, you won't be able to do so until after you create a user account and select the Bluetooth option from the desktop Settings app. And if you encounter any problems booting your Pi, you won't be able to interact with it from the command line since the Bluetooth connection might not be established. Using a dedicated wireless keyboard/mouse dongle communicates with the Pi just like any other wired USB keyboard and mouse. This assumes that the wireless keyboard you're using is fully charged, powered on, and successfully paired to the USB dongle that you would plug into one of the Pi's available USB ports.

If you plan to use your Pi as your primary desktop, or prefer to write all your code directly on the Pi, stick with a wired keyboard and mouse. If you intend to use the Pi as a media player or a portable data-collection device like I do, I recommend using the Corsair K83[5] wireless entertainment keyboard, shown in the next photo.

---

5. https://www.corsair.com/us/en/Categories/Products/Gaming-Keyboards/K83-Wireless-Entertainment-Keyboard/
p/CH-9268046-NA

In addition to its ability to interact with the Pi regardless of the state of the desktop, the K83 keyboard also minimizes clutter by not having a tangled mess of wires to unravel whenever you want to type or mouse click the Pi's desktop. Corsair's K83 keyboard is also backlit and has a large touch pad and mouse buttons, a built-in joystick, and a volume roller. It can connect to the Pi either via the preferred wireless USB dongle or via Bluetooth.

Water Sensor

With the standard peripherals like mice, keyboards, and displays accounted for, we can now turn our attention toward the sensors and network devices that will allow the Pi to do some pretty cool tasks for us. The least expensive of the sensors we'll employ is for Chapter 4, Water Leak Notifier, on page 49. You can purchase this sensor from Amazon[6] for around $13 US. Here's a photo of the sensor.

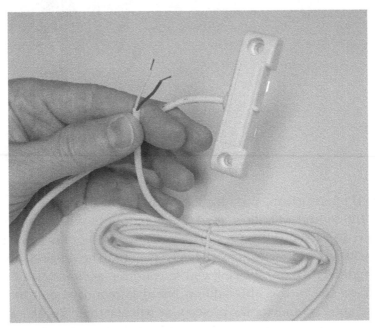

DockerPi SensorHub

Other projects in the book require specific sensors including motion detection, light reading, and temperature reading. We could purchase a dedicated sensor for each project, but that would quickly start filling up available GPIO (general-purpose input/output) pin connections on the Pi.

---

6. https://www.amazon.com/Floor-Water-Sensor-Flood-Detection/dp/B079YB1T8J?th=1

Fortunately for us, a certain Pi HAT (hardware attached on top) combines these and other sensors onto a single expansion board. Take a look at the following photo of the SensorHub and the external temperature probe attachment that comes with it.

The DockerPi SensorHub, available for around $22 US from Seeed,[7] mounts on top of the Pi and will allow our Python scripts to interrogate measured values from the SensorHub's five onboard sensors without needing to clutter up the Pi with wires attached to various GPIO pins. The SensorHub includes pass-through GPIO pins, allowing you to continue accessing and using the GPIO pins on the Pi that this add-on board is mounted on. The onboard sensors are good enough for the projects we will use them for, but if you need more precise readings, consider using a dedicated sensor for whatever specific variable you want to accurately measure.

Hue Starter Kit

If you don't already have a Philips Hue bridge in your home or office to automatically control lighting and power, consider purchasing the Hue Starter

---

7. https://www.seeedstudio.com/DockerPi-Sensor-Hub-Development-Board-p-4101.html

Kit.[8] You can purchase a hub and two white smart lightbulbs for around $70 US. The next photo shows, from left to right, a Hue bridge, smart bulb, and smart plug.

We'll be calling upon the Hue bridge in several projects, so if you just want a single smart light to start with, you can purchase the Hue bridge[9] and smart white light LED bulb[10] separately to save about $10 US.

In addition to turning on and off smart lights, we would also like to turn dumb appliances into smart ones. This can be achieved with the Hue smart plug[11] for around $30 US.

Simply plug in the smart plug into an available electrical socket, and plug the item to be powered on or off into the socket on the smart plug. We'll use the smart plug in Chapter 5 to turn on a desk fan when a room temperature exceeds a set threshold and turn the fan back off when the temperature drops below that desired threshold.

Smart Voice Assistant

The proliferation of voice-enabled assistants has brought the web to a conversational level. Smart assistants like Amazon Alexa and Assistant dominate

---

8.   https://www.philips-hue.com/en-us/products/smart-lighting-starter-kits
9.   https://www.philips-hue.com/en-us/p/hue-bridge/046677458478
10. https://www.philips-hue.com/en-us/p/hue-white-1-pack-e26/046677555689
11. https://www.philips-hue.com/en-us/p/hue-smart-plug/046677552343

the consumer voice-activated query market. Due to its low cost and remarkable voice recognition capabilities, the Google Home Mini device is one of the most popular voice assistants available today.You can use the Home Mini to play back everything from songs and news. It can also report weather forecasts and provide answers to how old a certain celebrity might be. Given its versatility, the Home Mini is also surprisingly inexpensive for a smart device. Here's a picture of the Google Home Mini.

Besides the water sensor, the Google Home Mini is the least expensive accessory we'll need for our projects. You can buy a new Home Mini from sellers on eBay[12] for under $20 US. Note that the various incarnations of Amazon Alexa hardware can also be used instead of the Google Home for these projects, but I'll be using the Home as my primary smart assistant in the book.

USB Infrared Transceiver

While the infrared control is becoming an ancient technology relic like the fax machine, a number of devices still rely on inexpensive line-of-sight IR transmissions to remotely control the appliance. I have an old HDTV display that still works great, but it isn't network-aware. Given all the hoopla around unwarranted data collection in modern smart TVs, perhaps that's a good thing. While I can use the old remote control that came with the TV, I programmed my Pi to turn it on for me. I'll show you how you can remotely control your infrared-enabled audiovisual equipment using this approach in Chapter 9, Voice IR Control, on page 119.

---

12. https://www.ebay.com/itm/Google-Home-Mini-Smart-Assistant-Charcoal-GA00216-US/114220854200

Since we're using a couple of GPIO pins for our other sensor projects and we'll likely use more in future designs, we can instead employ one of the available USB ports on the Pi to attach a USB-connected IR transceiver. Here's a picture of one such product.

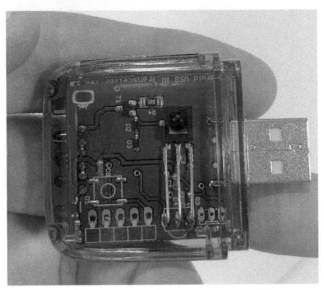

Designed and sold by Irdroid, this USB IR transceiver[13] can be used to both send and receive infrared signals. This can be used to not only transmit instruction codes to your IR appliances, but also to record codes being emitted by your own IR remote controls. You can even use a remote control to wirelessly communicate with your Pi. The Irdroid USB transceiver can be purchased for around $32 US.

Pi Camera Module V2

The last item on our project peripheral shopping list is the Raspberry Pi's camera module. The camera will allow us to capture high-resolution photos and stream video. While we could certainly use a standard USB webcam for this requirement, that would take up another valuable USB slot on the Pi, and such a USB webcam would likely be larger than the Pi itself. Just take a look at how small the Pi Camera V2 is in the photo on page 14.

The 8-megapixel V2 is the second iteration of the official Raspberry Pi camera, and it can be connected directly to the Pi's CSI (camera serial interface) via the camera's ribbon cable. We'll go over attaching and testing the Pi Camera Module V2 in Chapter 11, PhotoHook, on page 141.

---

13. https://www.irdroid.com/irdroid-usb-ir-transceiver/

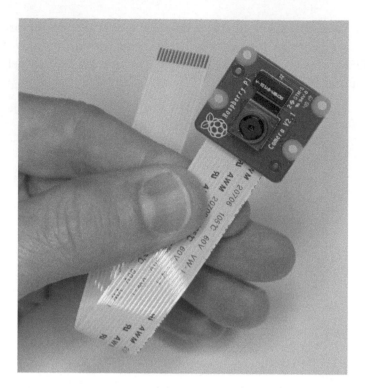

Next Steps

Assuming that you've acquired, at the very minimum, a Pi (a 4 Model B with at least 4 GB preferred), SD card with at least 32 GB storage, HD display with the necessary cable and adapter, mouse, keyboard, and USB-C power supply, you're ready to begin with installing the Raspberry Pi operating system. Roll up your sleeves, organize and connect your hardware layout, and prepare for an adventure in software installation, configuration, and testing in the next chapter.

CHAPTER 2

Setting Up the Software

With your Pi in hand along with the necessary peripherals, including power supply, SD card, monitor, mouse, and keyboard, you're ready to install and configure the operating system that will run on the hardware. Setting up a Pi has come a long way since the early days of the first generation Pi hardware. Even preparing the operating system on an SD card has gone from a multi-stage, multiple-application process to a simple three-button mouse click procedure.

By the end of this chapter, you'll have your Pi running the latest Raspberry Pi operating system with all the software needed to build the projects in the book. You'll also have an adequate computing experience to use the Pi as a passable, albeit sluggish, desktop replacement.

Raspberry Pi OS

The Raspberry Pi Foundation offers a Linux distribution installation tool for the Pi platform called the Raspberry Pi Imager.[1] Available for Apple macOS, Microsoft Windows, and Ubuntu Linux, the tool can image an SD card with your choice of OS distribution via just a few mouse clicks.

Download and run the Raspberry Pi Imager, and insert the SD card you want to image. Keep in mind that the inserted SD card will be completely reformatted, so if the card has any data on it that you want to keep, back it up before proceeding. The screenshot on page 16 shows what the Raspberry Pi Imager main screen looks like.

Selecting the Choose OS button will display a list of the recommended Linux distributions available for the Pi. While you can opt to install a more modern 64-bit operating system like Ubuntu Desktop, many of the Pi's hardware and

---

1.   https://www.raspberrypi.org/software/

software libraries we'll call upon in the book have not yet been ported to 64-bit. For hardcore hacker-types with plenty of time and patience, there are ways to shoehorn these libraries to work within these 64-bit platforms. At the time of this book's publication, a 64-bit alpha version of Raspberry Pi OS was in the works, but it was too unstable and incomplete to recommend, let alone build reliable projects upon. The next screenshot shows the different operating systems that the Imager can write to a target microSD card.

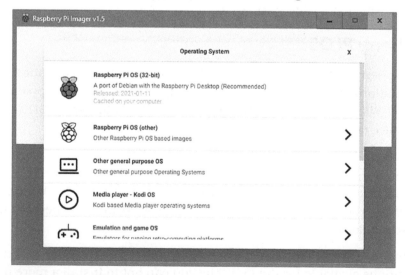

When you click the Choose OS button on the main Imager app screen, Raspberry Pi OS (32-bit) is displayed at the top of the list. Select that OS to continue.

With your SD card inserted and mounted on your host Windows, Mac, or Linux desktop computer, click the Choose SD Card button. When you're ready to commit the inserted SD card to be erased, reformatted, and installed with the Raspberry Pi operating system, select the Imager's Write button. The imaging writing and verification process will take about fifteen minutes or more to complete. When done, the Imager will announce you can remove the SD card from your computer and insert it into your Pi, as shown in the next screenshot.

Insert the microSD card into the slot underneath the Pi, making sure to properly align the card with its metal contact leads facing the Pi's board. Here's what the microSD card will look like when inserted into the bottom of a Pi 4.

Once properly seated, attach a keyboard and mouse to the Pi's available USB ports and connect an HDMI monitor to the Pi's micro-HDMI port via the micro-HDMI to HDMI adapter. Connect your Pi to its USB-C power source, and if the imaging was successful, you should see on the attached monitor your Pi boot into the Raspberry Pi OS desktop.

Once the Pi is up and running, it's time to answer a couple onscreen configuration questions in order for the operating system to know who you are, where you're located, and how it should connect to the Internet.

Configuring the Settings

Several dialogs will ask you to create a password for your Pi account as well as enter your network access, keyboard layout, and your geographic location. Assuming you have a wireless network in your home or office, connect to the network using your wireless network passphrase. We'll further explore the network settings needed to complete the initial installation configuration.

Networking

Modern Pi hardware offers both wired Ethernet and wireless network connectivity. If you're lucky enough to have completely wired your home or office with CAT-6 Ethernet cable, you can connect that cable to the Pi's Gigibit Ethernet port. If not, you can opt to connect wirelessly. Regardless of which network type you have, Raspberry Pi OS will prompt you through the process of connecting to your wireless network. Simply select your SSID identifier (in other words, the name of your wireless network) and connect with your wireless network's password. Once connected, the OS will eventually prompt you to download and install any security updates and patches that may have been released since the OS image was created.

When the latest updates are installed, you'll most likely need to reboot the Pi.

Finding the Pi on Your Network

The two most prevalent network configurations for computers and devices on your network are Dynamic Host Configuration Protocol (DHCP) and Static IP address assignment. DHCP is the most common default since it allows you to easily reconnect to other networks, such as bringing your Pi over to a friend's house and connecting to their network. This is because of the D in DHCP. That is, this protocol dynamically assigns the Pi's IP address based on a pool of IP's managed by your network router. While DHCP is convenient for client connectivity when you don't really care to know what IP address it uses, determining the actual IP address assigned to that device is different on an iPhone, Android, PC, Mac, or Pi. For example, issuing an ifconfig -a command in a Pi Terminal window will show details related to active network interfaces on the device, while an iPhone's graphic interface shows WiFi IP details via the Settings -> WiFi screen and selecting the Information icon. Additionally, the DHCP-assigned IP address may change depending on when the device was last powered on and connected to your network.

We could reduce this hassle by configuring the Pi's network interface to always use the same IP every time it's turned on. However, storing such Static IP assignments on the device makes it less portable, since you have to reassign the IP configuration every time you connect the Pi to another network. If you don't plan on ever changing your home or office network configuration and also don't plan on connecting your Pi to any other network but the one it was originally set up to use, then changing the network settings to Static IP assignment is acceptable. But what I've found much easier to manage is to use my home network router to automatically assign the same IP address to connected devices based on their media access control (MAC) address.

MAC Assignments and Macs

MAC assignments don't work by default on newer Apple products unless explicitly allowed by the user. This is because some networks use MAC addresses as a means of access authentication. MAC addresses can be easily tricked and tracked by bad guys. That's why using MAC assignments as a security mechanism is not a good idea. Apple uses a technique known as MAC Address Randomization to create new unique MACs on the fly. This makes IP Assignment based on MAC for Apple products moot unless the user intentionally disables MAC Address Randomization on their Apple devices. However, since we're only using the Pi's MAC as a means of assigning a network configuration and not for authentication, that should minimize any compromised security concerns on a home network.

Check your network router's ability to assign IP addresses based on device MAC addresses. Every router manufacturer does this in their own way. For example, on my Google Home WiFi setup, I have to go into the Advanced networking settings and select the DHCP IP Reservations to locate and add my Pi. Should the Pi ever go offline for an extended duration or connect to another DHCP-based network, I don't have to worry about the Pi acquiring a different IP address from my router since it will always be assigned the same one set in the router's IP Reservations. Doing so will also make it much easier to securely remotely connect to the Pi, as well as store and execute new scripts on it, because it will always have the same IP address assigned to it.

SSH Keys

With the OS fully patched and updated, you should now be able to safely launch the pre-installed Chromium web browser and surf the web. If all you want to use your Pi for is a capable, albeit sluggish, PC replacement, then you've accomplished your objective. But given how much more flexible the Pi is beyond a Chromebook, it's time to give the Pi the ability to act as both a client and a server. The quickest and most secure way to do this is to activate an OpenSSH server on the Pi. SSH stands for Secure Shell, and OpenSSH is the open source implementation of this remote access protocol. Using SSH will allow you to remotely connect and transfer files to the Pi from another computer on your network.

To install the OpenSSH server, open a Terminal window and enter the following installation command (don't enter the $ symbol, as that is simply illustrating the prompt indicator shown in the Pi's Terminal window):

```
$ sudo raspi-config
```

raspi-config is Raspberry Pi OS's built-in hardware and software configurator. Rather than manually editing boot configuration files, raspi-config allows you to easily activate hardware ports and software like OpenSSH. We'll revisit other raspi-config settings like camera and I2C interface access, but for now we'll enable SSH on the Pi by selecting Interface Options from the main screen.

On the next screen, select the P2 SSH option to Enable/disable remote command line access using SSH. In addition to enabling SSH access via the Terminal, this service will also allow you to transfer files to and from the Pi securely.

Once activated, you should be able to SSH into your Pi from another computer on your network that has the SSH client installed. If you're using a Mac or Windows 10 computer, an SSH client is included with those operating systems. To test your SSH access to your Pi, open a Command or PowerShell window on Windows or launch the Terminal app on macOS and enter the following command at the respective window's command prompt (substitute the IP_ADDRESS label with the IP Address of the Pi on your network):

```
$ ssh pi@IP_ADDRESS
```

As an example, my Pi has has an IP address assignment of 192.168.1.22. Therefore, to log in to my Pi from my Windows 10 computer, I open a Power-Shell window and type this in:

```
$ ssh pi@192.168.1.22
```

You'll then be prompted to type in your password that you created for your Pi during the initial OS setup procedure. When you SSH from another computer, you'll need to enter this password each time until you've set up digital keys for your SSH access. We'll do that shortly.

The first time you attempt to connect to your Pi via SSH, you'll receive a warning stating that the authenticity of the host computer you're attempting to SSH into can't be established. That's because you have never connected to the Pi from the other computer on your network that you are SSH'ing from. Since you know this is your Pi, you can type yes to the "Are you sure you want to continue connecting" question.

Now that you're connected to the Pi via SSH, you can access the file system, perform commands, and execute programs as if you were typing into the keyboard locally connected to your Pi. For example, typing the command ls at the Pi's command prompt will display a listing of the visible files and folders in your user's home directory on the Pi. If you're not familiar with Linux-centric command-line statements and utilities, check out the Linux.org[2] website for a number of well-written beginner tutorials.

Generating an SSH Key Pair

Now that you can connect to your Pi via SSH, it sure would be convenient if you didn't have to keep entering your password each time. To enable this, you can generate a unique set of digital keys that will allow your computer to connect to the Pi right away, no password required. This will also come in handy when we start using source control tools and code editors that may require these digital keys to already be in place.

Generating these keys takes a few steps but is fairly straightforward once you've done it a few times. You can enter the key generating command on your Windows 10 or macOS computer's Command/Terminal window:

```
$ ssh-keygen
```

Issuing this command will invoke the SSH key pair generating program, asking you where to save your digital keys. Accept the default location it suggests. The program will then ask if you want to protect your generated keys with an optional password. This is useful when you want an extra layer of security to protect your newly generated keys, but doing so will prompt you to enter that passphrase when you invoke their use. Since we want to

---

2. https://www.linux.org/forums/linux-beginner-tutorials.123/

access the Pi without having to enter a password each time we remotely log in to it, simply leave this passphrase blank.

Configuring the SSH Key Pair

The ssh-keygen program will create a .ssh directory in your Windows 10 or macOS user home directory and store two files, id_rsa and id_rsa.pub, in the .ssh directory. The id_rsa is your private key that is your computer's unique SSH identifier. The id_rsa.pub file is your public key that you will copy to your Pi's .ssh directory. To do so, first change to your Windows 10 or macOS .ssh directory within a PowerShell/Terminal window and then issue the following command (replacing the USERNAME and IP_ADDRESS values with your Pi's username and IP address):

```
$ cat ~/.ssh/id_rsa.pub | ssh USERNAME@IP_ADDRESS 'mkdir -p ~/.ssh && cat >> ~/.ssh/authorized_keys'
```

This string of commands will read the contents of your id_rsa.pub public key, SSH into your Pi, create a .ssh directory in your Pi's user home directory if it doesn't already exist, and then save your id_rsa.pub key details in a file in the Pi's .ssh directory called authorized_keys. The authorized_keys file is what the SSH server inspects whenever a key pair verification request is made in an SSH session. Think of the authorized_keys file as a door lock and your id_rsa private key file as the key to unlock the door. Protect and store the id_rsa file just as you would a real-life key, since anyone with a copy of that file could use it to gain remote access to your Pi.

If you configured the SSH key pair successfully, you should now be able to SSH into your Pi from your Windows or macOS computer without requiring a password. This time-saving security improvement will come in handy when we start developing our projects.

Python

The version of Python that comes pre-installed on Raspberry Pi OS is the latest stable version of Python 3. To determine which version of Python is installed on your Pi distribution, open the Terminal window and type python3 --version. Our projects will work with anything higher than Python 3.5, though Python 3.7 or higher is preferred.

For example, on the version of Raspberry Pi OS running on my Pi 4, the version of Python running on it is 3.7.3:

```
$ python3 --version
Python 3.7.3
```

Be sure to run python3, not python. The single python refers to the older Python 2.7 release that's included in the distribution to support older application dependencies that have not yet updated their libraries to the Python 3 interpreter.

The Python 3 distribution also includes Python's most popular package management utility called Pip. Using Pip makes it simple to download and set up additional helpful libraries that your Python scripts can call upon for extended functionality.

Pip Pip Hooray

As we begin to develop our projects, we'll be calling upon a number of additional Python packages not included in the core Python distribution. This is one of the reasons why Python is so powerful. Many smart and generous Python developers have already done the hard work of writing the code to access a number of web services, perform powerful data manipulations, interact with databases, and more.

The most popular Python package management and distribution system used today is called Pip. Pip is a recursive acronym for "Pip installs packages" and makes Python package installation a breeze. Pip3 is already installed on the Desktop version of Raspberry Pi OS. If you opted for the non-desktop version (Raspberry Pi OS Lite), you can install Pip for Python 3 using the apt install command:

```
$ sudo apt install python3-pip
```

A number of Python development libraries and build tools will be installed so that Pip can compile whatever packages may be requested. Installing these dependencies may take awhile, based on the speed of your Internet connection. Also, just like we execute python3 to launch the Python 3 version of the Python interpreter, we want to be especially sure to run the matching pip3 version of Pip when installing Python 3 packages. Otherwise, if we run just pip, we'll be installing packages for Python 2.7 instead:

```
$ pip3 --version
pip 18.1 from /usr/lib/python3/dist-packages/pip (python 3.7)
```

```
$ pip --version
pip 18.1 from /usr/lib/python2.7/dist-packages/pip (python 2.7)
```

Using pip3, you'll be able to retrieve and use the expanding world of Python packages. But be aware that each time you install a new package, you may also be installing a slew of additional third-party library dependencies along with it. As your Python 3 package installations grow, so too do potential library version conflicts, bloated and unwieldy system management, and performance issues. That's where the concept of virtual environments comes to the rescue.

Installing Pipenv

Several Python virtual environment utilities and configurations exist, but the one I've found that consistently does the job with the least amount of fuss is called Pipenv.[3] Pipenv combines the most useful Python virtual environment tools in a single, easy-to-use command-line utility. If you plan on creating many discrete Python projects running on your Pi, Pipenv will go a long way toward keeping Python dependencies isolated and well organized. While the projects in this book don't require you to use Pipenv, it's nevertheless an excellent tool with which to contain your Python project code.

Installing Pipenv is easy. To do so for all users, simply sudo the pip3 command in the Terminal.

```
$ sudo pip3 install pipenv
```

Once Pipenv is installed, I encourage you to review the variety of options this powerful, easy-to-use Python virtual environment manager has to offer.

By virtualizing your Python development environment, you can create discretely quarantined Python instances that have their own set of packages. You can even create virtual environments running different versions of the Python interpreter. This can come in handy when you want to test your Python code on different versions of Python without having to uninstall and reinstall those versions each time you want to run a test.

With the ability to install additional Python libraries, and the option to virtualize and containerize project instances, we still need a way to incrementally manage Python script source code. That is where Git comes in extremely handy.

Git

Git is the de facto source control solution for the technology industry and is the system we'll use for maintaining our project source code. Learning all the ins and outs of Git takes time and practice, but we'll focus on the most basic aspects of what Git has to offer, mainly creating new repositories and safely managing our changes to files saved to those repositories.

---

3.   https://pipenv.pypa.io/en/latest/

Virtual Environments, Containers, and Machines

Configuring software is a pain. This chapter is an example of how much time and effort is required just to configure a working environment, let alone the software dependencies needed to execute a script or application. One problem that Python coders encounter early on is the buildup of third-party libraries used for their projects. Some of the more sophisticated libraries depend on dozens of additional libraries that need to be installed, and some of these require specific versions of such library dependencies. If you were to install all these libraries in the same global bucket, conflicts would inevitably arise. It can also create potential security risks if an untrusted source is carrying a malicious payload or has access to administrative system privileges. For example, it could run a bad actor script during its unauthorized execution and install unwanted and even dangerous programs that could steal your passwords, CPU cycles, or be used in attacking other computer systems. These are just some of the reasons why creating virtual environments for your Python scripts is a good idea. You could even go further with this approach by using application containers and full-blown dedicated virtual machines.

The 8 GB model of the Pi 4 has enough RAM to run a virtual machine, but its processor simply isn't the best CPU for running VMs. Your high-end desktop PC or dedicated server hardware will host VMs far better. While VMs provide an excellent degree of OS and application dependency isolation, doing so for a simple script is a huge resource overhead. Even on an 8 GB Pi, you would run out of system resources after running only a few VMs. That's where application containers like Docker[a] come in handy. Let's say you wrote a cool home-automation script that requires a long list of library packages and you wanted to share that script with others. You could have a detailed list of instructions for your collaborators to follow, or you could save them a lot of time by building a predefined image that already includes all the dependencies and configurations required. All they would need to do is download the image and run it. If you plan on collaborating with others on a Python project that requires a lot of dependency configuration, Docker is a helpful and free Pi-capable tool that could save you a lot of time and sanity in the long run.

---

a. https://www.docker.com/

Take it from me, it pays to expend the extra effort to create repositories and manage your project's code using this defacto source control tool. Nothing is worse than realizing you accidentally overwrote hours, days, or even months of work while tinkering with source code. When used correctly and consistently, Git will not only be able to recover from those accidents but also allow you to safely collaborate with other developers working on the same project source as you.

Like Python and Pip, Git is already pre-installed on Raspberry Pi OS. If you opted instead to run the Raspberry Pi OS Lite version, you can install Git using the apt install command:

```
$ sudo apt install git
```

Once installed, you need to personalize Git so the utility knows who you are when you save file updates to Git repositories. Do so by entering the following two instructions in the Terminal window, replacing your name and email address in the quotations indicated:

```
$ git config --global user.name "Your First and Last Name"
$ git config --global user.email "your@emailaddress.com"
```

Verify that your name and email address were saved correctly. The following Terminal window command should return your first and last name.

```
$ git config user.name
```

Entering the following command should return the email address you entered in the previous global user.email statement.

```
$ git config user.email
```

If you need to modify either of these identifiers, simply rerun the git config --global command for the appropriate field.

 Joe asks:

Can't I just use a cloud service like GitHub to host my Git repositories instead of locally running my own Git server?

GitHub is an excellent resource for hosting Git repositories. But the biggest mistake new developers make is posting their code containing sensitive passwords and other confidential details on GitHub. Countless email accounts, VPNs, and company passwords have been inadvertently compromised this way.

It's best to host your own Git repositories while gaining proficiency using Git commands and features. Also review more advanced Python practices that show how to import sensitive data like passwords and API keys as environment variables stored in a separate, secure file. When you're ready to share your code with others and have removed any confidential details from it, you can easily graduate to posting your projects on GitHub or other repo hosting services.

Now that Git is tailored for your use, we'll create individual repositories for each of the book's projects when we begin work on them. For now, create an empty repository to practice the basic Git commands of storing, retrieving, and copying files in a Git repository. We'll begin with the commands used in a Terminal window, but for our projects, we will use an easy graphical tool to store and update project code changes.

In preparation for our first project, let's create a repository to store the project's code and any changes we make to it. First, use the File Manager app (by selecting the icon in the upper-left corner of the Pi OS desktop that looks like two yellow file folders) and create a folder called repository in your Home directory, as shown in the next screenshot.

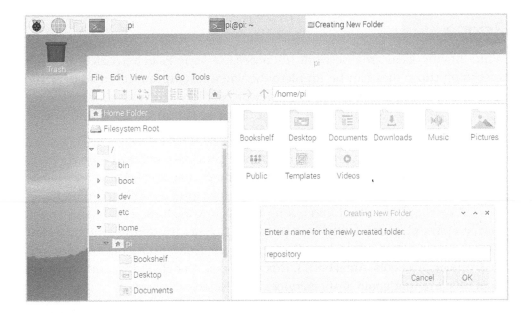

Next, open a Terminal window and enter the following Git command:

```
$ git init --bare repository/twitchtimer.git
```

If successful, Git will report that it initialized an empty Git repository in the requested folder path. We'll use this new repository to store and manage the Twitch Timer project code.

Joe asks:
Where can I learn more about how to use Git?

Git is a very powerful source control system that has many more features and commands beyond the simple clone, add, commit, and push commands used in our projects. As your code becomes more sophisticated and you want to try out different ideas, you can create additional branches to your files to test without affecting the main source files. Branching comes in handy when you have other people contributing to your project. You can learn more about Git by visiting the Git website.[a]

---

a.   https://git-scm.com/doc

Rclone

While having local area network file access to your Pi is extremely helpful, accessing those files can be problematic when attempting to do so over the Internet. You could set up a VPN server or port forward your connection requests from your router to your Pi, but doing so is cumbersome and potentially insecure.

The ever-growing popularity of cloud storage providers makes using them on platforms like Android, iOS, macOS, and Windows easy and broadly accessible. But few of these providers offer a simple app for Linux to participate. Rclone,[4] created by Nic Craig-Wood and a long list of contributors, is a free and open source cloud platform utility written in the Go[5] programming language that allows users to connect to over thirty different cloud storage providers and file transfer protocols. Amazon S3, Box, Dropbox, Google Drive, and Microsoft OneDrive are just some of the services that Rclone supports.

Because Rclone is a command-line utility, setting it up is not as easy as running a nice vendor-supplied GUI installer. But what it lacks in pretty dialog boxes, it excels in tremendous power and flexibility. The following steps will connect to and mount a Google Drive to the Pi user's home directory. If you prefer to use Dropbox or OneDrive instead, simply select those services instead.

Install Rclone using the typical apt install command:

```
$ sudo apt install rclone
```

---

4.   https://rclone.org/
5.   https://golang.org/

Configuring Rclone

Configure Rclone to work with your cloud storage provider of choice:

```
$ rclone config
```

In the initial configuration screen that follows, type n to create a new remote and assign that new remote a relevant name. Since I'm using Google Drive, I called mine GDrive.

The Rclone configuration process will display a numbered list of cloud storage providers and ask you to enter the number of the storage provider you want to use.

Select Google Drive from the list. Rclone will ask if there is an application client ID associated with this connection. Normally this is blank, so proceed accordingly. Rclone then asks if there's an application client secret. This is blank by default, and pressing the keyboard's Enter key will submit that acceptable default value. Finally, Rclone asks what scope of access you would prefer to have to your account's cloud-stored files. If you prefer to have all your Google Drive files accessible on the Pi, choose the first option (1), Full access all files, excluding Application Data folder.

If required, you can then enter the ID of the root folder. This is normally blank by default, and you can simply press the Enter key to continue with the configuration. Similarly, you can elect to have a noninteractive login, but once again, entering a blank value is the default and works for our purposes.

Rclone then asks if you would like to edit advanced configurations. This isn't necessary for our basic needs, so just choose n for No. Then it asks if you want to use auto config, steering you to answer y for Yes unless you have a nonstandard setup scenario. We don't, so select y.

Now, Rclone is ready to connect to your Google Drive and will launch the Chromium browser on your Pi desktop to complete the OAUTH key exchange. Enter your Google login credentials, and then click the Allow button to allow Rclone to access your Google Account.

If successful, the browser will report Success! All done. Please go back to rclone. You may now close the browser and return to the Terminal window to finish up.

Rclone next asks if you want to configure this connection as a team drive. For this single-use purpose, you can respond with n for No.

Finally, Rclone will ask if the configuration is OK before committing the configuration of the Google Drive access on your Pi. At this point, you can add more cloud service providers to your Pi, but you should be fine with the configured Google Drive for now. Select q to exit the Rclone configuration routine.

Rclone Folder Assignment

Now that we have Rclone able to connect to our Google Drive, we need to make it easy to see the files in the desktop File Manager.

Launch the File Manager and create a new folder called GoogleDrive in the home directory, as shown in the next screenshot.

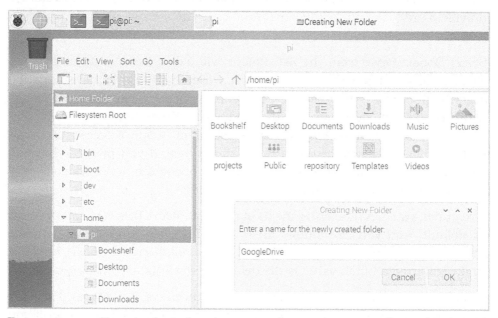

Return to your Terminal window to map and mount your configured Rclone'd Google Drive to the newly created 'GoogleDrive' folder using the following command:

```
$ rclone --vfs-cache-mode writes mount GDrive: ~/GoogleDrive
```

This command tells Rclone to allow cached writes and mount the Google Drive to the GoogleDrive folder that was created. Assuming everything was entered correctly, you should now see a Google Drive icon appear on your desktop.

Opening that drive should reveal the files and folders stored in the cloud on your Google Drive.

That's pretty neat! But watch what happens when you close that Terminal window running the rclone command. The Google Drive icon on the desktop vanishes along with the connection to your Google Drive files and folders. To make the mounted GoogleDrive persistent on our desktop even when we reboot the Pi, we need to add that lengthy rclone command as an executable shell script and then tell the OS to automatically run that script when your Pi account logs into the desktop.

Launch Rclone on Boot

Begin by opening a text editor like Nano in the Terminal window and add the rclone command for mounting the GDrive that we used earlier:

```
$ rclone --vfs-cache-mode writes mount GDrive: ~/GoogleDrive
```

After typing or pasting in the rclone mount instructions into Nano, press the Ctrl+O keys to save the file. Name this new file mountgdrive.sh. Exit Nano via Ctrl+X to return to the command line in your open Terminal window. Then turn the mountgdrive.sh file into an executable via the chmod command:

```
$ chmod +x mountgdrive.sh
```

Test this now executable script by opening the desktop File Manager and double-clicking the mountgdrive.sh file to run it. Note that the icon represented for this script marked as executable is represented by a sprocket. When you double-click to run the script, File Manager will ask if you want to Execute, Execute in Terminal, Open, or Cancel your action, as shown in the next screenshot.

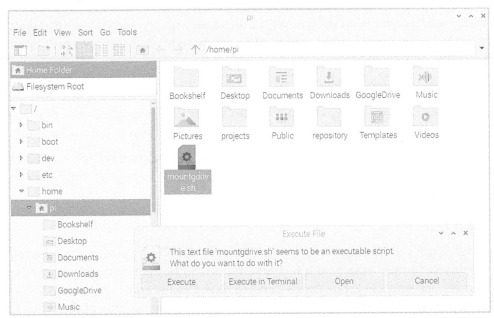

Selecting Execute will launch the script without opening a separate Terminal window. Execute in Terminal will do the same but first open a Terminal window from which the script will execute. Open simply opens the script file in the default desktop text editor (typically called Mousepad in the standard Raspberry Pi OS distribution), in case you need to make further edits. Since we just want to execute the script without showing another Terminal window, choose the Execute button option. Assuming all goes well, your Google Drive should mount and display once again on the Pi's desktop.

Note that even if you log out and log back in to your Pi desktop, Google Drive is still mounted. That's nice, since you only have to manually execute the script once to persist the mounted drive. But if you reboot, you have to manually rerun the script to remount the drive.

Let's complete this setup by automatically running the mountgdrive.sh shell script when the desktop launches. To do so, we need to create a crontab. A crontab is a program that uses a specially formatted text file to run scripts or applications at specific intervals. In our case, we'll create an entry in the crontab to run the mountgdrive.sh script whenever the desktop launches. Open a Terminal window and start the crontab program and pass it the -e argument to indicate you want to edit your crontab file:

```
$ crontab -e
```

If this is the first time you've run crontab, it will ask you which text editor you prefer to use to edit your crontab file. Since we're already comfortable with Nano, choose that option. With the crontab file open, move your cursor via the down arrow on the keyboard to the bottom of the screen and enter the following command:

```
@reboot /home/pi/mountgdrive.sh &
```

This command instructs crontab to run the script as soon as the desktop launches. The & character at the end tells the script to run in the background.

Enter Ctrl+O to write out the modified crontab file and then Ctrl+X to close Nano. You can learn more about crontab at any time by typing crontab --help in the Terminal window.

Try It Out

Alright, time to test all our work. Reboot your Pi. When you see the desktop displayed, it may take a couple additional seconds to show the icons in the launch bar and desktop. That's because our rclone script is running, and it takes several seconds to establish the connection and mounting of your Google Drive. Also note that unlike before, a Google Drive icon doesn't display on the desktop. That's fine, since we can still access the content of our Google Drive via the File Explorer app. Simply navigate to the GoogleDrive folder and viola, your Google Drive files are there. You can now save and retrieve files to your Google Drive with ease. Additionally, we'll be using this established functionality in Chapter 7, PiSpeak, on page 83.

It's unfortunate that the major cloud storage players like Google and Microsoft don't offer an easier installer for the Pi and Linux in general. I suspect it's not because they don't want the Pi to participate in their ecosystems, but rather it's due to the broad variations of desktop overlays that Linux users can choose from. From a macOS or Windows perspective, it's always the same graphic desktop shell. Linux offers choice, but with choice comes complexity. If you were able to automatically mount your Google Drive successfully after a reboot, then you've successfully navigated and overcome this complexity and should commend yourself accordingly.

Take a break at this point if you need to, and enjoy a job well done. Otherwise, press onward with setting up a supervisor and installing a polished, powerful, yet free commercial desktop text editor.

Supervisor

As we create several different projects throughout the book, the scripts we write will need to keep running even if we're not sitting in front of the Pi waiting for events to occur. Normally when you run a Python script from a Terminal, the Python script will stop working when you close that Terminal window or session.

You can instruct a script to continue to run in the background even after the terminal from which it was started is closed. However, there's no guarantee that such a script will always continue to run, especially if there are memory or other resource constraints. Additionally, it would be useful if we could have certain project scripts automatically start running when the Pi reboots. What we need is a supervisor that will launch those scripts, keep them running, and restart them if they should happen to stop working. Supervisor[6] is the solution that will do exactly what we need. We can install Supervisor using the apt install command:

```
$ sudo apt install supervisor
```

This will install the older version of Supervisor installed as a system-level application. A newer release of Supervisor can be installed via the Pip3 command but requires additional tooling to get it working reliably as a system service. Regardless of which version you choose, both will be able to manage

---

6. http://supervisord.org/

Python 2.7 or Python 3 scripts. Confirm that the Supervisor package was deployed successfully by checking the version of the supervisord daemon that is running on your Pi

```
$ supervisord --version
3.3.5
```

We'll explore Supervisor's features later and configure it for our needs when we deploy Python scripts throughout the book, beginning in Chapter 4, Water Leak Notifier, on page 49.

Code Editor

I prefer using graphical text editors over character-based terminal editors, such as GNU nano[7] or Vim.[8] A competent text editor optimized for writing and executing Python code, called Thonny Python IDE, comes pre-installed on Raspberry Pi OS. This editor is perfectly fine for entering and running the Python scripts in this book, but if you want to use a world-class editor with numerous options and expansive add-ons, consider Microsoft's free Visual Studio Code editor.

TextMate[9] on macOS was the gold standard before Sublime Text[10] became my top choice for several years. However, I switched to Microsoft's free Visual Studio Code editor several months ago and haven't looked back. Visual Studio Code offers nearly every base feature I need and has grown even more popular than other editors due to its ever-expanding library of free extensions. And unlike TextMate or Sublime Text, VS Code also runs natively on the Pi desktop!

Installing Visual Studio Code on the Pi

You can install Visual Studio Code on a Pi in several ways. The easiest is to simply install it using the apt install command.

```
$ sudo apt install code
```

You can find the Visual Studio Code launch icon in the Programming menu, just below the Thonny Python IDE.

If you've never used a visual text editor before, using Visual Studio Code can be intimidating at first. Even so, I find it far easier than using a terminal-based text editor, mostly due to all the time-saving shortcuts and extensibility

---

7. https://www.nano-editor.org/
8. https://www.vim.org/
9. https://macromates.com/
10. https://www.sublimetext.com/

that VS Code has to offer. Feel free to play around with the menu options, window panes, and icons in the editor to get more familiar with the program. It will come in handy for some of the longer Python scripts we'll write later in the book.

Installing Visual Studio Code Extensions

Microsoft's Python extension is one of the most helpful and popular extensions available for Visual Studio Code. It provides an array of useful Python programming enhancements, including syntax highlighting, automatic code formatting, virtual environment support, and more. Using the Python extension will help considerably with writing and testing the Python code used in the book's projects.

Install the Python extension by selecting the Extensions option from the main View menu. This will display a listing of the most popular extensions available. It's no surprise that the Python extension is at the top of the list. To install and activate this extension, select the Install button on the accompanying Python description page.

I have another significant recommendation regarding using VS Code on the macOS or Windows platform. Recall that we set up the SSH keys to allow easy remote login access to the Pi from the Terminal window. VS Code has a useful, time-saving free extension called the Visual Studio Code Remote - SSH: Editing Configuration Files Extension,[11] as shown in the next image.

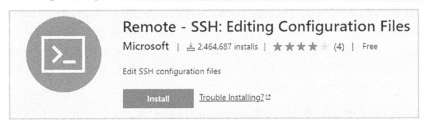

This extension will give you the ability to remotely create and modify files on your Pi from your Mac, Windows, or Linux laptop or desktop computer. This ability comes in handy when your Pi has been deployed somewhere in your home or office that isn't convenient to access, or when you no longer have or need peripherals like a mouse, keyboard, and monitor attached to your Pi while it's running in the background. When you decide to further explore the family of Pi devices, such as the Pi Zero, you'll discover that having such convenient access to a remote file system via VS Code's Remote SSH extension is a sanity saver.

---

11. https://marketplace.visualstudio.com/items?itemName=ms-vscode-remote.remote-ssh-edit

Since you've already successfully configured and tested SSH access and related keys on your PC or Mac, adding the SSH extension to the Mac or PC version of VS Code is as simple as launching VS Code on that desired platform (assuming of course that you've already installed Visual Studio Code on your PC or Mac computer), selecting Extensions from VS Code's View menu, and typing in "SSH editing" into the extension search field. The Visual Studio Code Remote - SSH: Editing Configuration Files Extension[12] should appear at the top of the list. Clicking the Install button on that extension's description page will install the extension into your desktop or laptop's running version of VS Code.

To connect and edit files on your Pi, select the Remote Explorer icon from VS Code's left-margin toolbar, then choose SSH Targets from the Remote Explorer drop-down selector. Add your Pi to the list by choosing the + icon next to the SSH Targets drop-down indicator. Then type in your account username and the IP address (pi@192.168.1.22 as an example) of your Pi in the pop-up text box.

Once connected, you'll be able to remotely traverse and open your Pi's text files directly within VS Code running on your desktop or laptop with ease.

Next Steps

If you were able to successfully configure all the recommended software and settings in this chapter, then your Pi is ready for all the projects in this book. If you encountered any issues setting up your Pi following these instructions, be sure to follow the steps exactly as indicated. Even skipping one minor detail can railroad an entire setup. If you're still having problems with any of the steps even after several attempts and web searches, post your experiences on the Pragmatic Bookshelf forum so that I and other astute readers of this book can do our best to assist you.

Assuming everything went well and your Pi is connected and working as expected, take a victory break before diving into the book's projects. Spend time poking around the Raspberry Pi OS desktop, surf the web, and launch a few of the default applications to get an idea of the responsiveness of the environment. When you're ready to get back into the flow, proceed with the Twitch Timer project. That project will show you how easy it is to watch your favorite Twitch stream on your Pi without the need of a web browser, a custom Twitch application, or even a Twitch login account.

---

12. https://marketplace.visualstudio.com/items?itemName=ms-vscode-remote.remote-ssh-edit

Part II

Projects

With the necessary hardware and software acquired and ready to go, we can proceed with using these components in the range of projects in this part of the book.

CHAPTER 3

Twitch Timer

One of the more popular uses for the Raspberry Pi is converting it into a media center. Pi-optimized OS distributions, such as LibreELEC[1] and OSMC (Open Source Media Center),[2] give the Pi an attractive and easy way to navigate to and access open source and commercial media services, such as YouTube, Netflix, and Twitch. If all you want to use your Pi for is to consume streaming content, these options do an amazing job. But if you want to use your Pi for more than just media consumption and have full control over what you run on your hardware, this project shows one way to solve this problem.

I enjoy watching my favorite Twitch streamers after a long day of work, usually settling in bed as the streams play on. But too many times I woke up in the middle of the night with my display on and Twitch unwatched streams still consuming bandwidth. We will discuss one way to turn off the TV with our Pi in Chapter 9, Voice IR Control, on page 119, so this project will focus entirely on how we can use our Pi to connect to Twitch, direct playback to the popular VLC player, and then turn off the stream after a certain duration of time has passed. And we will do all of this with just a few short lines of Python code.

Setup

Here's what you need to build this project.

Hardware

- HDTV display, or computer monitor with HDMI input and built-in speakers

---

1. https://libreelec.tv/
2. https://osmc.tv/

Software

- VLC media player (already installed in Raspberry Pi OS Desktop distribution)
- Streamlink[3]

The VLC media player comes pre-installed with Raspberry Pi OS Desktop. VLC is one of the most popular open source media players available due to its ability to play back an expansive range of different media formats. One of those media formats is the file type that Twitch uses to stream their content.

Locate the VLC icon (it looks like an orange-and-white striped safety cone) in the Pi's Sound & Video desktop menu and launch the application. Right now, it's just a black screen with a larger version of the application's icon in the center and not that interesting to look at. Let's change that. Close the app and follow along.

We'll use the Streamlink package for this project. But before we do, let's create a projects folder to store the book's various Python project scripts. Open a Terminal window and enter the mkdir command:

```
$ mkdir ~/projects/
```

Then, using the cd command, change to that directory so you can clone a Git repo into it:

```
$ cd ~/projects/
```

Recall from the last chapter that you already created a Git repository for this project in advance. Connect to and synchronize with that Git repository now:

```
$ git clone ~/repository/twitchtimer.git
```

The git clone command will create a new folder called twitchtimer within the projects folder. Change to the twitchtimer folder using the cd command:

```
$ cd ~/projects/twitchtimer
```

Now let's create our initial Python script file using the touch command and commit that initially empty file to the repository:

```
$ touch twitchtimer.py
$ git add twitchtimer.py
$ git commit -m "Initial TwitchTimer Project commit."
$ git push
```

---

3. https://streamlink.github.io/

The touch command is used to create an empty file, in this case, twitchtimer.py. This will be the file we will use to save our Python code for this project. The next command, git add, adds files to our local Git store in the twitchtimer folder that we cloned from our Git repository. The command git commit with the additional -m "Initial TwitchTimer Project commit." argument commits and permanently stores the changes we made to this file since the last time the commit command was run. The -m indicates the optional message we want to include with our commit. These messages can be helpful when looking back on prior commits for a particular change made to a file or set of files.

Finally, the git push command by default attempts to push our committed changes to the original Git repository that we cloned. In this case, it will replicate the changes we made in the twitchtimer directory to the ~/repository/twitchtimer.git Git repository. We'll rerun these git commands each time we make an important or noteworthy change to the twitchtimer.py file. We'll use this same cyclical git approach for each subsequent project in the book.

Next, install the Python-based Streamlink application using pip3 in a Terminal window:

```
$ sudo pip3 install streamlink
```

Running this command will install Streamlink and the additional third-party Python libraries it depends on to function correctly. You can check to see which version of Streamlink was installed by asking for its version in the Terminal:

```
$ streamlink --version
streamlink 2.0.0
```

Streamlink is a powerful and flexible command-line utility that gives users the ability to redirect media streams like those used by Twitch and YouTube to media players like VLC. Due to its flexibility, the number of arguments capable of being passed to the streamlink command can be daunting. You can learn more about the variety of parameters that can accompany the streamlink commmand by reading the program's User Guide.[4] While there are literally dozens of parameters that can be used, we'll focus on specific Twitch arguments for this project.

To verify that both VLC and Streamlink have been installed successfully, connect to a Twitch stream that broadcasts twenty-four hours a day. One of those is the MST3K Twitch channel. To determine the name of a Twitch

---

4.   https://streamlink.github.io/

channel, launch a web browser and navigate to twitch.com. Enter mst3k in the Search box on Twitch's home page. The search results should show MST3K at the top of the page. Select MST3K and note the URL, which now reads twitch.tv/mst3k. Therefore, mst3k is the name of the MST3K channel. Try several other searches to get comfortable identifying the channel name this way.

Now that we know MST3K's channel name on Twitch, we can pass that into Streamlink as a command-line argument along with the quality of video stream playback:

```
$ streamlink twitch.tv/mst3k 720p
```

If everything was installed and configured correctly, the VLC player will launch and play the MST3K live Twitch stream.

Note that not all Twitch streams broadcast in 720p, while others that do may opt to broadcast their 720p streams at sixty frames per second (720p60). To verify the available formats and resolutions that the stream uses, run the streamlink command without indicating the stream quality. In the case of the MST3K stream, doing so produces the following results:

```
$ streamlink twitch.tv/mst3k
Available streams: audio_only, 160p (worst), 360p, 480p, 720p, 1080p (best)
```

Depending on the Pi you have and the bandwidth of your Internet connection, you may see the stream playback quality occasionally stutter. If so, try reducing the quality of the stream from 720p resolution to 480p:

```
$ streamlink twitch.tv/mst3k 480p
```

MST3K was originally broadcast in the 90s before the advent of HDTV, so 480p should look no different compared to higher resolutions. In addition to being a lighter load on the Pi's central and graphics processors, 480p will also consume less bandwidth on your network.

Now that we have Twitch playback in VLC, let's write a simple Python script that will automatically launch this stream and turn it off after a certain amount of time has passed.

Creating a Sleep-Timer Script

If you installed Visual Studio Code, you're welcome to use that editor to open the empty twitchtimer.py file we created earlier. But given the brevity of this project's Python script, it may be easier to use the Nano text editor for now:

```
$ nano twitchtimer.py
```

Either manually enter or copy and paste the following Python code into the editor window:

```
twitchtimer/twitchtest.py
```

```python
① import os, subprocess, signal, sys, time

② duration = int(sys.argv[1])*60

③ cmd = "streamlink twitch.tv/mst3k 720p"
④ twitchstream = subprocess.Popen(cmd,shell=True)

⑤ time.sleep(duration)
⑥ os.killpg(os.getpgid(twitchstream.pid), signal.SIGTERM)
```

❶ The first line of code imports the Python standard libraries we'll depend upon in later parts of the script.

❷ We'll pass a numeric value as an argument when we run our script. This integer will indicate the number of minutes we want to stream before forcing the stream to quit. Python's time.sleep() method will pause script execution for the number of seconds asssigned to it. Since the argument being passed in is minutes, we need to convert that value to seconds by multiplying it by 60.

❸ We'll create a string variable called cmd that will store the command we want our Python script to execute. In this case, it's the same command we used to initial test our Streamlink installation.

❹ We'll create a subprocess within our Python script that will execute the streamlink command, connect to the MST3K livestream, and open the VLC media player application to view the stream.

❺ We'll instruct the Python script to pause further execution of any remaining script until the number of minutes we passed in the script's argument have passed.

❻ Once the number of seconds stored in the duration variable have passed, we'll tell our Python script to kill the processes it created when it launched the streamlink command. This includes any subprocesses it created, such as quitting any VLC player that was launched.

Save the file and return to the Terminal window. Run the script and pass it the number 2, like this:

```
$ python3 twitchtimer.py 2
```

The number 2 will be multiplied by 60, and store the result, 120 seconds, to the script's duration variable. That's how long the script will wait after it launches the Streamlink app. After the two minutes (120 seconds) have passed, the

script will kill the Streamlink process along with the VLC program playing
the stream:

```
$ python3 twitchtimer.py 2
```

Once successfully tested, commit your recently saved twitchtimer.py Python
script to your Git repository:

```
git commit -m "Initial Python script using streamlink process call."
```

TwitchTimer Script Improvements

We're essentially done with our project since we've successfully connected to
a broadcasting Twitch channel, opened VLC, and closed VLC and the stream
after a set duration of time. But our script isn't very robust. For example, try
running the script without passing a numeric value (or even no value at all).
It will fail to run, reporting either a ValueError or IndexError depending on what
was or wasn't passed as an argument.

Also, while MST3K provides a continuous stream of entertainment on their
Twitch channel, it isn't the only broadcaster on Twitch. It would be convenient
if we could pass another argument to our script indicating which channel we'd
like to watch for how long. Fortunately, we're not the only ones who need robust
command-line processing in our Python script. argparse[5] is a built-in Python
library we can import into the twitchtimer.py script to help parse our script's
command-line arguments. We can also use it to inform users of our script
what kind of arguments it requires to execute the script via helpful instructions
and an example. This will come in handy months after you wrote such a script
but forgot which arguments the script requires in order to properly execute.

Let's modify our original script to include the argparse library and apply it
toward the duration and channel arguments to be passed. We'll also include
some basic error handling to more gracefully handle any exceptions that may
arise. Lastly, as a bonus, we'll kill the process differently depending on which
operating system this script is executed on:

twitchtimer/twitchtimer.py
```python
import argparse, os, platform, signal, subprocess, sys, time

parser = argparse.ArgumentParser()
parser.add_argument('-c', action='store', dest='channel',
help='Twitch name of channel to stream', required=True)
parser.add_argument('-t', action='store', dest='duration',
help='Number of minutes you wish to stream', required=True, type=int)
```

---

5. https://docs.python.org/3/library/argparse.html

```
        args = parser.parse_args()

        duration = int(args.duration)*60
❷   try:
                cmd = "streamlink twitch.tv/" + args.channel + " 720p"
                twitchstream = subprocess.Popen(cmd,shell=True, stdout=subprocess.PIPE)
❸           while True:
                        line = str(twitchstream.stdout.readline())
                        if not line:
                                break
                        if line.__contains__('error: Unable to validate '):
                                print('Unable to connect to channel, either because \
                                channel is not broadcasting now or channel name is \
                                invalid. Quitting.')
                                quit()
                        if line.__contains__('Opening stream'):
                                break

                print('Stream ' + args.channel + ' for ' + str(duration) + ' seconds.')
❹           time.sleep(duration)
❺           if platform.system() == 'Windows':
                        os.system("taskkill /IM vlc.exe")
                else:
                        os.killpg(os.getpgid(twitchstream.pid), signal.SIGTERM)

                print('Streamed ' + args.channel + ' for ' + \
                 str(args.duration) + ' minutes.')
❻           twitchstream.kill()
                time.sleep(5)
                os._exit(-1)
                time.sleep(5)
                sys.exit(-1)
    finally:
                print('Stream closed.')
```

❶ After importing the argparse library, we can use it to not only parse the
command-line parameters passed into our script but also add some help
text and even simple validation to confirm the type of variables we're
passing. And to eliminate any ambiguity on the position of the values
we're passing, we'll add the requirement of a channel to be preceded with
a -c flag and the playback time with a -t flag. Since we can't process the
script without either of these values, we'll make them required. This way,
if one or both values are missing from the command line, our script will
gracefully fail with instructions for the user on what values it needs to
proceed.

❷ Whenever a script executes a routine or, as in our case, an external program that could have unexpected or unplanned results, it's always a good idea to wrap such routines in a try/catch/finally block. If an unexpected error does occur, the script can attempt to recover from the error. Several errors can occur in our script, such as a broken pipe (an unexpected loss of a data stream). Normally we would try to identify each of these error conditions and instruct our script to recover from these errors with whatever corrective code needs to be executed. But in the interest of time and space, we short-circuited this comprehensive approach in favor of using the finally keyword to act as a catchall bucket should an error occur. In this case, when we do kill the VLC player and end the stream, if we don't properly wait for the subprocess to exit gracefully, we can print out a Stream ended. message in the Terminal.

❸ The While True loop checks to see if the channel name we passed to streamlink is a valid, currently broadcasting channel. If streamlink fails to connect to a requested channel's stream, we'll gracefully print the result to the Terminal. If streamlink is able to connect, thereby reporting that it is Opening stream, we'll exit the loop and proceed with the remainder of the script's execution.

❹ Just as in our original script, we pause further execution of the script for the duration passed in our command-line argument.

❺ This entirely optional condition checks to see if we're executing this script on Windows. If so, Windows handles process termination differently than Linux. The only reason I included this check is because this is the one project that doesn't depend on Pi-specific hardware to execute. Python is portable for a reason, and this script exemplifies Python's ability to run on multiple OSs and hardware platforms. If you wanted to run this script on a Windows laptop, for example, you could do so as long as Streamlink and VLC player were installed on it.

❻ Add a few extra timers and system calls to account for the number of Streamlink subprocesses being unexpectedly closed when we terminate the VLC player. If you commented out these commands via the # character and ran the script, you would likely see a slew of errors appear in your Terminal when the script terminates. The time.sleep(5) function waits five seconds before calling the os._exit(-1) and sys.exit(-1) functions. Doing so accomplishes the objective of gracefully quitting the stream and closing the player.

Try running the script, explicitly passing the -c and -t values. For example, if we want the MST3K channel to play for two minutes as a test, execute the following in your Terminal window:

```
$ python3 twitchtimer.py -c mst3k -t 2
```

As long as the script is correctly entered and formatted with the proper indents, you should see the VLC player appear and play the stream for two minutes before quitting the player and returning the playback channel and duration in the Terminal window. To demonstrate the portability of this script, here's an example of Terminal window output after a successfully concluded playback session on a Windows computer:

```
python3 .\twitch.py -c mst3k -t 2
Stream mst3k for 120 seconds.
SUCCESS: Sent termination signal to the process "vlc.exe" with PID 15920.
Streamed mst3k for 2 minutes.
```

Try different combinations of channels and durations to see how the script handles them. Try omitting either value or flag to test the argument error handling. Enter a bogus channel name to see how it recovers from "streamlink unable to connect to the stream." Also, if the script appears to hang, make sure the streaming resolution value is set to one that the channel supports, as was mentioned earlier in this chapter.

Once you're satisfied with the script's revisions, remember to commit and push the changes to your Git repository:

```
$ git add .
```

```
$ git commit -m "Improved argument handler and exception handling."
```

```
$ git push
```

Congratulations on completing the first project in the book. In an effort to minimize editorial redundancy, future projects won't explicitly review steps for setting up the Git for those projects.

Next Steps

The Twitch Timer demonstrates how easy it is to turn your Pi into a streaming media playback device with a programmable sleep timer. Now you can fall asleep and not worry about waking up hours later to an unwatched stream and gigabytes of unnecessarily wasted bandwidth.

Upon completing a later project in Chapter 9, Voice IR Control, on page 119, you could add the ability to turn off the TV monitor you were watching the stream on, assuming the TV uses an infrared (IR) remote control to turn the TV on or off. You could also stop the requested Twitch stream after a set duration.

You could also start up the Twitch Timer by reacting to Discord[6] or email notification, alerting you that your favorite Twitch streamer is live. But for now, sit back and enjoy the progress you made in this chapter by watching a few minutes of your preferred Twitch channel.

---

6.   https://discord.com/

CHAPTER 4

Water Leak Notifier

As demonstrated in our previous project, the Pi can be used as an inexpensive laptop or desktop replacement, capable of running Python scripts that can also run on other operating systems. What makes the Pi so interesting in the home automation world is its diminutive size and expansive ports and GPIO pins. Sensors and actuators can be attached to the Pi via these interfaces, giving the Pi a wide spectrum of inputs to process. One of the simplest of inputs is a switch.

For this project, we'll use a very basic water sensor that turns on (completes or closes an electrical circuit) when water is detected across its two probes and turns off (breaks or opens an electrical circuit) when water is no longer present across both probes. To make this switch more useful, we'll write a Python script that not only detects if the switch is on or off but also sends an email when water is detected.

Setup

Here's what you need to build this project.

Hardware

- Floor water sensor[1]

Software

- Gmail (or other IMAP-compliant email) account

Attach one of the water sensor's white wires to pin 39, which is associated with ground on the Pi 4, and the sensor's red wire to pin 40, which is associated with GPIO 21. GPIO 21 (pin 40) provides the power source and the ground

---

1. https://www.amazon.com/Floor-Water-Sensor-Flood-Detection/dp/B079YB1T8J?th=1

GPIOs and Pi Models

The GPIO Pinout map is different depending on which model of Raspberry Pi you are using. This book assumes you have the Pi 4 Model B+ and will reference these GPIO pin locations accordingly. Here's a useful tip from one of this book's technical reviewers and fellow Pragmatic Bookshelf author Maik Schmidt: if you're running Raspberry Pi OS, you can type in pinout in the Terminal window, and a graphical map and list of all the pinouts for the model of the Pi that you have will be displayed onscreen. Very cool!

pin is the drain. When looking at a properly oriented Pi 4, pins 39 and 40 are the furthest-right bottom and top pins, respectively.

Since the water sensor doesn't have female adapters for the Raspberry Pi's GPIO pins, and since it's not a good idea to directly solder the wires directly to the Pi's GPIO pins, I recommend using female to female jumper wires. You can affix the water sensor wires into one end of the jumper wires and then hold it in place either using electrical tape or, my personal favorite, heat-shrink tubing.

If you opt to seal the female jumper wire with the exposed water sensor wiring, use a heat-generating source such as a hair dryer on high setting to shrink the tubing around the connection.

Whenever water is detected across both of the water sensor's probes, the electrical circuit will close, meaning the switch is flipped on. When this occurs, GPIO 21 should report that it senses current flowing through it. Refer to the photo on page 51 to see where the water sensor wires should attach to the Pi GPIO pins.

With the male leads of the two water sensor wires converted into female ends, you can easily attach and remove the connection. This will come in handy when adding additional sensors or other attachments that you want to test, either in conjunction with or isolated from the water sensor connection.

Test Script

Before we flesh out a full-blown Python script to handle our email notification workflow, let's first test the water sensor to be certain that it is working as expected.

We'll be calling upon a custom Raspberry Pi library called RPi.GPIO to interrogate the status of the GPIO pins. Like many other Pi-centric/Pi-optimized software, the RPi.GPIO Python library is already pre-installed in Raspberry Pi OS. So there's no need to use Pip to install it.

Create an empty Python file called watersensortest.py. Open the file using your preferred text editor and enter the following Python code:

watersensor/watersensortest.py

```
① import RPi.GPIO as GPIO
   import time

② GPIO.setmode(GPIO.BCM)
   GPIO.setup(21, GPIO.IN, pull_up_down = GPIO.PUD_UP)

   alert_trigger = False

③ while True:
       if GPIO.input(21):
```

```
④         if alert_trigger != True:
              print("No water detected.")
⑤             alert_trigger = True
          else:
              if alert_trigger != False:
                  print("Water detected.")
                  alert_trigger = False
          time.sleep(1)
```

❶ Import the RPi.GPIO library that we'll use to poll the status of GPIO pin 21.

❷ Initialize the GPIO library and set the status of pin 21 to UP via the GPIO.setup parameter pull_up_down = GPIO.PUD_UP.

❸ Run an infinite loop to poll the status of GPIO 21 every second.

❹ If GPIO 21 is pulled down (that is, water is detected and the circuit is closed), then print that water has been detected. Otherwise print no water detected.

❺ To prevent a constant stream of water/no water messages being printed to the Terminal window, set a variable called alert_trigger to True or False depending on whether the state change message has been outputted already to the Terminal window.

Save and execute the script with python3:

```
$ python3 watersensortest.py
```

If everything is coded error free and your water sensor is working and correctly connected to your Pi, you should see the output of a single line, No water detected. Dip the two metal probes of the water sensor into a small cup of water and check your Terminal window. A new Water detected. message should appear. Remove the sensor from the water and another No water detected. message should be displayed.

You can continue to repeat this test as many times as you prefer. Press the CTRL+C keys on your keyboard when you're satisfied with the accuracy and consistency of the results to interrupt and escape the forever looping While routine in the script.

Water Sensor Email Alerts

While having the sensor print out the status of the water sensor to the Terminal window is useful, it would be impractical for us to sit in front of the Terminal all day long waiting for water to be detected. It would be more helpful to receive an email notification whenever we have a water leak from a dishwasher, clothes washing machine, or even a basement leak. Using the print statement stubs we already have in place from our test script, we can easily add an email function that can be called in lieu of printing.

If you have your own email server already set up, you can use that server instead of public servers. But nearly everyone has a Gmail account, so this example will use Gmail as the mail server that will relay the water alerts to you. Note that you can set up a similar configuration on other public email services like Microsoft's Outlook.com or Yahoo Mail.

Since we'll be storing our email account passwords in an unencrypted, plain-text Python script, I strongly advise that you create a brand-new, free Gmail account to use exclusively for your Pi and home automation email notification projects. Should your Python scripts ever get compromised for any reason, you limit your mail storage exposure to simple notification messages.

Now that we've created a dedicated Gmail account and configured it to allow messages to be sent and received from other mail clients such as a Python script, let's create a brief emailtest.py script to verify that our new email address is working:

watersensor/emailtest.py

```
❶ import smtplib
   import email.mime.multipart
   from email.mime.text import MIMEText

❷ def send_email(message):
❸     gmail_user = 'YOUR-GMAIL-ACCOUNT-NAME@gmail.com'
       gmail_password = 'YOUR-GMAIL-ACCOUNT-PASSWORD'

       try:
❹         msg = email.mime.multipart.MIMEMultipart()
           msg['to'] = 'Your Name<YOUR-EMAIL-ADDRESS@DOMAIN.COM>'
           msg['from'] = 'Your Gmail<YOUR-GMAIL-ACCOUNT-NAME@gmail.com>'
           msg['subject'] = 'Email Test'
           msg.add_header('reply-to', 'YOUR-GMAIL-ACCOUNT-NAME@gmail.com')
           msg.attach(MIMEText(message, 'plain'))
           session = smtplib.SMTP('smtp.gmail.com', 587)
           session.starttls()
           session.login(gmail_user, gmail_password)
           message = msg.as_string()
           session.sendmail('YOUR-GMAIL-ACCOUNT-NAME@gmail.com',
           'YOUR-EMAIL-ADDRESS@DOMAIN.COM', message)
           session.quit()
           print('Email sent.')
       except:
           print('Email failed to send.')

❺ send_email('This is the body of the Email Test message.')
```

Let's quickly review the code used in this script before we attempt to run it.

❶ Use and import the built-in smtp and email Python libraries.

❷ Create a send_mail() function that accepts a message string to appear in the body of the email.

❸ Replace the example gmail_user and gmail_password assignments with the Gmail username and password credentials from your own Gmail account. And no, you can't use the one I have in the example since it's bogus anyway.

❹ Set up the necessary MIMEMultipart() code. You could have used a more basic plain-text mailer for your messages, but it's better to set up the multipart MIME option up front to easily allow you to add HTML-formatted message bodies in emails, should you decide to do so in the future.

❺ If the script is able to authenticate to Gmail and successfully send your email, you'll see Email sent. displayed in your Terminal window after running the script. If there's an authentication or configuration error when the script is executed, you will alternatively see Email failed to send. in your Terminal.

Save and run the script:

```
$ python3 emailtest.py
```

The first time you run your script, you will be blocked by Gmail and you'll see Email failed to send. in your Terminal window. That's because Gmail has increased security on their Gmail accounts due to the rise in fradulent Gmail logins and abuse. However, since we've intentionally created a new Gmail account only for use with sending notification emails from our Python scripts, log in to your new Gmail account and there should be an email alert from Google indicating that a login attempt was made and Google blocked it. Verify that the login attempt was actually from your script by noting the time, date, and location that Google indicated on the alert. Follow the link provided in the email, then Allow Less secure app access within your new Gmail account.

Run your script again, and as long as your username and password are correct and accepted by Google, you should see an Email sent. message in your Terminal window. Checking the email Inbox of the email you sent the test message to should show that you received the

Email Test

PB P4 Book <p4book@gmail.com>
    3:11 PM

To: Mike Riley

This is the body of the Email Test message.

test message. Here's a photo of one of the emails I received during my testing.

You can now send emails with customized subjects and body messages from a Python script. That's pretty cool. Now you can combine this newfound messaging capability with your water sensor test script to make a new watersensor.py script:

watersensor/watersensor.py

```
import RPi.GPIO as GPIO
import time
import smtplib
import email.mime.multipart
from email.mime.text import MIMEText

GPIO.setmode(GPIO.BCM)
GPIO.setup(21, GPIO.IN, pull_up_down = GPIO.PUD_UP)

def send_email(message):
    gmail_user = 'YOUR-GMAIL-ACCOUNT-NAME@gmail.com'
    gmail_password = 'YOUR-GMAIL-ACCOUNT-PASSWORD'

    try:
        msg = email.mime.multipart.MIMEMultipart()
        msg['to'] = 'Your Name<YOUR-EMAIL-ADDRESS@DOMAIN.COM>'
        msg['from'] = 'Your Gmail<YOUR-GMAIL-ACCOUNT-NAME@gmail.com>'
        msg['subject'] = 'Water Sensor Alert'
        msg.add_header('reply-to', 'YOUR-GMAIL-ACCOUNT-NAME@gmail.com')
        msg.attach(MIMEText(message, 'plain'))
        session = smtplib.SMTP('smtp.gmail.com', 587)
        session.starttls()
        session.login(gmail_user, gmail_password)
        message = msg.as_string()
        session.sendmail('YOUR-GMAIL-ACCOUNT-NAME@gmail.com',
        'YOUR-EMAIL-ADDRESS@DOMAIN.COM', message)
        session.quit()
        print('Email sent.')
    except:
        print('Email failed to send.')

alert_trigger = False

while True:
    if GPIO.input(21):
        if alert_trigger != True:
            send_email("No water detected.")
            alert_trigger = True
    else:
        if alert_trigger != False:
            send_email("Water detected.")
            alert_trigger = False
    time.sleep(1)
```

This script is simply a combination of the watersensortest.py and emailtest.py Python scripts, replacing the watersensortest.py print functions with the send_email()

function. Like before, run the newly combined script with a cup of water nearby to verify the receipt of water detection email messages:

```
$ python3 watersensor.py
```

As soon as you execute the script, you should receive an email sent to the address you supplied. Test that the email messages are being sent and received each time you dip the water sensor into a cup of water and pull it back out again. You should see the water detection status message in the body of each email. If you want to make these emails less verbose, you can modify the subject line to include the water status. For example, replace the line:

```
msg['subject'] = 'Water Sensor Alert'
```

with

```
msg['subject'] = 'Water Sensor Alert - ' + message
```

When you receive emails with this modified subject line, they'll show the water sensor status without having to open the email to view the message body. Later on when we attach the Pi Camera to the Raspberry Pi, you could improve this email notification by having the camera take a photo of the area being monitored and include it as an email attachment.

Sensor Supervision

Recall how back in Supervisor, on page 33, we installed the Supervisor utility. The time has come to put that daemon application to good use. While you certainly can just launch the watersensor.py script from a Terminal window, as soon as you close that window, the script will stop. If you recall the rclone script in Rclone, on page 28, one way to prevent that from occurring is to add an ampersand character, &, to the end of the command:

```
python watersensor.py &
```

This ampersand instructs the BASH shell to run the script in the background and to keep it running even if the Terminal window originating the script's execution is closed. While the script may continue to run in the background this way, there's no guarantee that it will continue to run. Resource constraints, conflicts, or even a reboot will kill the script and there's no way for you to know without testing to see if you still receive email alerts.

The Supervisor utility solves this conundrum by not only launching the script at boot if you so desire; it also constantly monitors the script to verify it is running. If the script should die for any reason, Supervisor will execute the script again. Setting up infinite looping scripts like watersensor.py this way gives us the peace of mind that the script will be running in the background even if the Pi accidentally or intentionally reboots. As long as the Pi is running, a correctly configured and error-free supervised script will be running too.

Configuring Supervisor to launch and monitor our script is mildly verbose but fairly easy, especially after you've configured the utility a few times. In the case of our watersensor.py script, we need to create a new configuration file in the /etc/supervisor/conf.d/ directory. This configuration file instructs Supervisor which Python interpreter to use, the Python script to run, which user to run it as, whether it should automatically run the script at boot time, whether it should automatically restart the script if it should unexpectedly die, and more. Let's create a new configuration file for our watersensor.py script using the Nano utility:

sudo nano /etc/supervisor/conf.d/watersensor.conf

Enter the following configuration settings in this newly created watersensor.conf file, replacing the YOURUSERNAME placeholder with the username of the acccount on your Pi that you want to execute the watersensor.py script. If you're using the default Pi user account, then replace YOURUSERNAME with pi:

```
watersensor/watersensor.conf
❶ [program:watersensor]
❷ directory = /home/YOURUSERNAME/projects/watersensor
❸ command = /usr/bin/nohup /usr/bin/python3 watersensor.py &
❹ user = YOURUSERNAME
  environment HOME="/home/YOURUSERNAME",USER="YOURUSERNAME"
❺ autostart = true
❻ autorestart = true
❼ stdout_logfile = /var/log/supervisor/watersensor.log
❽ stderr_logfile = /var/log/supervisor/watersensor_err.log
```

Once entered, save the configuration file in the /etc/supervisor/conf.d directory by pressing the CTRL+O keyboard keys. Now let's examine the configuration file more closely.

❶ This label is used to identify the name of the program that Supervisor will run.

❷ This is the directory that your watersensor.py script is currently located in.

❸ The nohup program is a built-in Linux app that allows a command or script to keep running even after you log out of your Pi. Python 3 is the intended Python interpreter we want to use to run the watersensor.py script. And finally, we explicitly run the /usr/bin/python3 Python 3 interpreter and pass the script name along with the ampersand character to instruct the script to run in the background. Note that it's good practice to provide absolute paths to the intended programs in your Supervisor configuration files. This guarantees that Supervisor will be able to locate and run the applications responsible for correctly executing your configured Python scripts.

❹ Set the user and home environment variables to your user account and absolute home directory path.

❺ We want the watersensor.py script to automatically begin running at boot time. If the Pi should reboot on its own or we need to reboot it due to a system software update, we don't have to remember to manually start the script on its own.

❻ We also want the watersensor.py script to automatically restart should it unexpectedly quit. Note that if you have a bug that crashes your script, Supervisor will continue to rerun the script even as it continues to repeatedly crash. This is why it's important to fully debug your script before setting up a Supervisor configuration file. Doing so will insure that Supervisor will keep the script running as error free as possible.

❼ If you have any print statements that would normally output to a Terminal window, these statements will be sent to a log file. Good Supervisor log file practice dictates that log files should be stored in the /var/log/supervisor directory for easy access and review. In this case, we'll save these outputs to the /var/log/supervisor/watersensor.log log file.

❽ Lastly, if your script does encounter an error and/or crashes, the errors emitted by such events can be reviewed in the stderr log file. This error log will be stored in the /var/log/supervisor/watersensor_err.log log file.

For Supervisor to utilize your newly created watersensor.conf configuration, you can restart the Supervisor service by entering the restart command in the Terminal window.

```
$ sudo service supervisor restart
```

Conversely, you can opt to reboot your Pi.

```
$ sudo reboot
```

It's best to execute the reboot command, so you can check to see if the script automatically executes once the Pi finishes rebooting. Verify that the script automatically launched and is running in the background by performing the water cup test on the water sensor. If you received the expected email alerts, you're good to go! If not, inspect the /var/log/supervisor/watersensor_err.log file to see what errors prevented the script from running or why the script might have crashed. Address the errors and restart/reboot and try again.

Once properly configured, the Supervisor program is a powerful utility that empowers the Pi with incredible autonomy and flexibility. We'll be calling upon Supervisor a few more times throughout the book as we create more projects that continuously poll the status of various sensors attached to the Pi.

Joe asks:
What about the Pi Zero?

While our powerful Pi 4 will diligently perform the job of polling the status of the attached water sensor, dedicating this hardware to such a simple, non-CPU-intensive task seems like a gross underutilization of the Pi 4's hardware capabilities. Of course there may be unique scenarios that may be able to multitask the Pi to check water status while streaming media, but that assumes a sensor with wire long enough to reach the testing source (not to mention keeping such wires from becoming a safety hazard).

Fortunately, for these kinds of low-CPU-powered projects that need a Linux OS, the Raspberry Pi Foundation has a special kind of Pi called the Pi Zero W.[a] This ten-dollar gumstick-sized computer has just enough compute capacity to execute projects such as these while still offering most of the features, such as WiFi and Bluetooth connectivity, found in its larger siblings. Take a look in the next photo at how small the Pi Zero is in my hand.

Due to the Pi Zero W's size, it takes a bit more effort to set up and attach wires to the GPIO headers (particularly since there are no pins arising from the board), but for small, single-purpose dedicated-network functionality, the Pi Zero W is an amazing piece of technology miniaturized into a tiny yet powerful networked computer. In fact, I use my Pi Zero as a backup Git server and an automated price tracker in between checking for water leaks.

a. https://www.raspberrypi.org/products/raspberry-pi-zero-w/

Great job on building a water sensing computer that emails you whenever a water leak is detected! You can use this newfound capability in any scenario involving a collection of water. In my case, I want to know whenever my washing machine starts spewing water onto the floor. This usually occurs when the lint filter gets clogged and results in a mess sometimes taking hours to mop up. With my water sensor in place, I no longer have to worry about my washing machine unexpectedly turning my basement into a swimming pool.

Next Steps

This project has shown you how easy it is to construct a fairly sophisticated, autonomous monitoring system. By doing so, you now have a good foundation to custom build and utilize other home automation ideas. You don't need to rely on third-party subscription services or foreign servers storing your home data without your knowledge. You are in full control and management of your own project data processes.

Our next project will apply the same concepts we used in this project. We'll use similar techniques to check the status of a different type of sensor and control a popular consumer lighting and electricity management system.

CHAPTER 5

Hue Fan

Being an avid tech enthusiast, I surround my tinkering workspace with running computers of all shapes and sizes. While these machines keep me warm with all their heat exhaust in the winter, they turn my office into a sauna during the hot summer months.

To keep me and my machines cool, I have a floor fan to help dissipate the heat. But I often forget to turn on the fan until after my sweat-drenched shirt sticks to the back of my chair. Likewise, I forget to turn the fan off when evening comes or the computers are powered down. To save energy while keeping me cool, my Pi acts as a thermostat to turn on or off the floor fan, based on the room temperature.

Building this solution is easy thanks to the code we already wrote in the previous chapter. All we need to do is change the type of sensor and make a call upon a home automation platform. Let's make something cool!

Setup

Here's what you need to build this project.

Hardware

- DockerPi SensorHub[1]
- Philips Hue bridge[2]
- Philips Hue smart plug[3]
- Desk fan or other appliance that can be toggled on or off via a power knob or switch.

---

1. https://www.seeedstudio.com/DockerPi-Sensor-Hub-Development-Board-p-4101.html
2. https://www.philips-hue.com/en-us/p/hue-bridge/046677458478
3. https://www.philips-hue.com/en-us/p/hue-smart-plug/046677552343

Software

- Hue Smartphone[4] app
- phue[5] Python library

Start by installing and configuring your Philips Hue bridge and Hue accessories, such as smart lights and smart plugs, by following the instructions in the Hue bridge package. Use the Philips Hue smartphone app to connect to, label, and organize your accessories. You will need these labels to identify the smart plug and lights to control from the Python scripts in this book. The Hue app provides an easy way to organize and assign names to your Hue lights and appliances.

Be sure to power off and unplug the Pi before attaching the SensorHub, as shown in the next photo.

Before you can query the data that the SensorHub is collecting, you must enable the Pi's ARM I2C kernel module. Do so via the raspi-config program:

```
$ sudo raspi-config
```

Select Option 3, Interface Options from the main menu, and then choose the P5 I2C menu option to Enable the automatic loading of I2C kernel module.

---

4. https://apps.apple.com/us/app/philips-hue/id1055281310
5. https://github.com/studioimaginaire/phue

Seating the Sensor

The SensorHub package includes four metal posts that can help secure the SensorHub more permanently to the Pi. I suggest you test your SensorHub attached to the Pi's GPIO pins first before securing it with the posts. When you're satisfied with the long-term use of the SensorHub, you can make the installation more secure by screwing the posts in place. Look at how much more secure the SensorHub sits on my Pi 4 in the next photo.

When you power on the Pi with the SensorHub correctly attached, the SensorHub's red LED power indicator located in the upper-left quadrant on top of the SensorHub will remain constantly illuminated. You may also see a blue LED near the middle-right edge of the SensorHub occasionally turn on and off. This LED lights up whenever motion is detected within the last five seconds. You can verify this by waving your hand over the board to watch this detection occur.

Since we'll be using the Philips Hue platform to turn on and off a Hue smart plug connected to a fan (or other appliance that may be better suited for your needs), we'll be calling upon the phue third-party Python library. Install the library into your virtual environment:

```
$ sudo pip3 install phue
```

Before we can start sending commands to the Hue bridge, we have to register it first.

Bridge Registration

When you installed the Hue system, you needed to press the sync button on the top of the Hue bridge for your Hue smartphone app to talk to the unit. Philips has adopted a basic means of security to prevent anyone nearby from gaining unauthorized access to your Hue network. By pressing the button on the Hue bridge, you are authorizing the application requesting control access.

To do this for Python scripts running on the Pi, you first need to register with the bridge by connecting to it within thirty seconds after the Hue bridge button has been pressed. You'll also need to know the IP address of the Hue bridge on your network, which you should be able to identify from your network router's dashboard, similar to the way the Pi's IP and MAC address were identified back in Finding the Pi on Your Network, on page 18.

Create the following script in your huefan directory, name it register.py, and add the lines of Hue bridge registration code to this file:

huefan/register.py
```
from phue import Bridge
huebridge = Bridge('YOUR_HUE_BRIDGE_IP_GOES_HERE')
huebridge.connect()
```

Let's take a quick look at what this three-line Python script does.

❶ Using the from Python keyword, we can import a specific class or function from a module and assign it a to an instance variable. In this case, we import Bridge from the phue module for use in our script.

❷ We pass the IP address of our Hue bridge to Bridge and assign its result to the huebridge variable.

❸ As long as the button on the top of the Hue bridge was pressed a few seconds before this script is run, the Hue bridge will register our Pi and grant it the ability to control other Hue-registered devices on the network. Once you've successfully registered your Pi with the Hue bridge, there's no reason to call the connect() again unless you use a different Hue bridge or another Pi to run your Hue scripts on.

Replace the YOUR_HUE_BRIDGE_IP_GOES_HERE with the IP address of your Hue bridge. When you're ready, press the sync button on top of your Hue bridge and to allow it to automatically register new devices asking to register with it. In our case, this new device is our Pi running this script.

Execute the register.py script within thirty seconds of pressing the Hue bridge button. If you fail to run the script in time, you will have to press the Hue bridge button and run the registration script again.

```
$ python3 register.py
```

If no output appears after you execute the script, then you successfully registered your Pi with the Hue bridge. You can now use your Pi and the phue scripts you run on it to remotely control other Hue devices on your network.

If the script failed to connect(), it will provide details in the output errors. The most frequent error for not being able to register is using an incorrect IP address for the Hue bridge. This error is usually reported at the bottom of the error trace as a socket.gaierror: [Errno -2] Name or service not known. Fix the IP address and connectivity problem so that you can successfully register your Pi with the Hue bridge before proceeding.

Registered Devices

To confirm your effort was successful as well as identify the name and ID of the connected Hue smart plug, create and run the following listdevices.py script that will connect to your Hue bridge and enumerate a list of Hue accessories configured with the bridge:

```
huefan/listdevices.py
from phue import Bridge
❶ huebridge = Bridge('YOUR_HUE_BRIDGE_IP_GOES_HERE')
❷ lights = huebridge.lights
❸ for light in lights:
      print(light.name)
```

Let's take a quick look at how this script connects to your Hue bridge and identifies the other Hue devices it's able to manage.

❶ Here we pass the IP address of our Hue bridge into the huebridge object instance.

❷ Next, we interrogate the bridge lights collection and assign those details about the Hue smart bulbs, smart plugs, and other Hue-controllable accessories to the lights variable.

❸ Finally, we use a for loop to iterate over the lights collection and assign each item to the light variable. Then we print the name we assigned each of those devices via the Hue smartphone app.

The output of the script will list the names of the Hue lights, smart plugs, and accessories you have registered and named with your Hue bridge. Identify the name of your Hue smart plug from the list. In my case, I named my smart plug "Fan", a name I will use in the sample code. If you labeled your smart plug or light a different name using the Philips Hue app on your smartphone, be sure to replace 'Fan' in the code listings with the name of your Hue-labeled device. For example, here's the output shown after running the python3 listdevices.py command from within my Terminal window.

Now that we see our list of connected devices, we can turn them on or off using a single line of Python code.

Turn It On Again

Create a smartplugtest.py file, and replace the YOUR_HUE_BRIDGE_IP_GOES_HERE and the Fan identifiers with the IP address and Hue name of the device in your environment:

huefan/smartplugtest.py
```
from phue import Bridge
import time

huebridge = Bridge('YOUR_HUE_BRIDGE_IP_GOES_HERE')
```
❶ `huebridge.set_light('Fan','on', True)`
```
time.sleep(10)
```
❷ `huebridge.set_light('Fan','on', False)`

❶ Power on the device by calling the set_light() function with the name of the device you want to control (in this case, Fan), and setting its value to True.

❷ After waiting ten seconds, the set_light() function is called again, this time setting the Fan device to False to turn it off.

Be near the fan or whatever other appliance you connected your smart plug to and verify that your smart plug is working before running the script. Do this using the Philips Hue app on your smartphone by toggling on and off the smart plug. Once confirmed that everything works as expected, set the smart plug to off via the Hue smartphone app and execute the smartplugtest.py script:

```
$ python3 smartplugtest.py
```

Your smart plug should power on. After ten seconds, it should power off. If you had a fan or other power toggle appliance connected to the smart plug, it should do the same.

Sensing the Heat

Now that you can power on Hue lights and plugs with a single line of Python code, let's combine this newfound capability with a temperature sensor. When the temperature meets or exceeds 27 degrees Celsius (roughly 80 degrees Fahrenheit), turn on the fan connected to the smart plug. When the temperature drops below 27 degrees, turn the fan off.

Create a file called temperaturetest.py and write code to display the temperature reading from the SensorHub's temperature probe once every second. The SensorHub initialization code was obtained from DockerPi's wiki:[6]

huefan/temperaturesensor.py
```
import smbus
import time

DEVICE_BUS = 1
DEVICE_ADDR = 0x17

TEMP_REG = 0x01
LIGHT_REG_L = 0x02
LIGHT_REG_H = 0x03
STATUS_REG = 0x04
ON_BOARD_TEMP_REG = 0x05
ON_BOARD_HUMIDITY_REG = 0x06
ON_BOARD_SENSOR_ERROR = 0x07
BMP280_TEMP_REG = 0x08
BMP280_PRESSURE_REG_L = 0x09
BMP280_PRESSURE_REG_M = 0x0A
```

---

6. https://wiki.52pi.com/index.php/DockerPi_Sensor_Hub_Development_Board_SKU:_EP-0106

```
BMP280_PRESSURE_REG_H = 0x0B
BMP280_STATUS = 0x0C
MOTION_DETECT = 0x0D

bus = smbus.SMBus(DEVICE_BUS)
```

❸
```
while True:
    aReceiveBuf = []

    aReceiveBuf.append(0x00)

    for i in range(TEMP_REG,MOTION_DETECT + 1):
        aReceiveBuf.append(bus.read_byte_data(DEVICE_ADDR, i))
```

❹
```
    if aReceiveBuf[STATUS_REG] & 0x01 :
        print('The external temperature sensor range was exceeded. \
        No value reported.')
    elif aReceiveBuf[STATUS_REG] & 0x02 :
        print('The external temperature sensor is disconnected.')
    else :
        external_celsius = aReceiveBuf[TEMP_REG]
        external_fahrenheit = (external_celsius * 9/5) + \
```
❺
```
        32
        print('The current external sensor temperature = '
        + str(external_celsius) + ' C / ' + \
        str(external_fahrenheit) + ' F\n')
```

❻
```
    onboard_celsius = aReceiveBuf[ON_BOARD_TEMP_REG]
    onboard_fahrenheit = (onboard_celsius * 9/5) + 32
```

```
    print('The current onboard sensor temperature = '
```
❼
```
    + str(onboard_celsius) + ' C / ' + str(onboard_fahrenheit) + ' F')
```

```
    if aReceiveBuf[ON_BOARD_SENSOR_ERROR] != 0 :
        print('The current onboard temperature and \
        humidity sensor data is unavailable.')

    time.sleep(1)
```

Let's review what this script is doing before executing it.

❶ To communicate with the I2C interface connecting the SensorHub to the Pi, we need to import the smbus Python library. This external library is already preloaded on the Raspberry Pi OS Desktop. If you encounter an error when running this script, be sure to have the smbus library installed (via `sudo pip3 install smbus`) and verify that the SensorHub is correctly mounted and operational on the Pi.

❷ This is the initialization and address assignment code required to communicate with the various sensors onboard the SensorHub. For this project, we'll be focusing on just the external and onboard temperature sensors. We'll call upon the SensorHub's motion and light sensors in Chapter 6,

Hue Auto Light, on page 75. A barometric pressure sensor is also on the SensorHub, but it won't be used in any of this book's projects.

❸ This While statement will always be True, and thus the script will run until you Ctrl+C escape out of the script's execution.

❹ As mentioned previously, the SensorHub has two temperature sensors: one onboard that is soldered directly on the board itself, and an external temperature probe that more accurately detects room temperature. Be sure to have the external sensor probe attached to the SensorHub before running the script. The conditional statement elif aReceiveBuf[STATUS_REG] & 0x02 will check if the probe is attached and report if it's not.

❺ The board reports its temperature in the Celsius scale. For those who prefer the Fahrenheit scale instead, we perform a quick conversion so we can print the temperature in both Celsius and Fahrenheit measurements.

❻ Even though we'll use the external temperature probe readings for this project, the onboard sensor reading is being reported just to illustrate the temperature variations between the two sensors. The onboard temperature sensor is literally sitting on top of the Pi's most heat-generating chips. As these chips heat up, so will the temperature around this sensor. In my experience, the onboard and external temperature sensor readings can vary well beyond 10 degrees Celsius (20 degrees Fahrenheit).

When you're ready, run the script:

```
$ python3 temperaturesensor.py
```

Note the temperature readings for both the onboard and external sensors being printed out every second in your Terminal window.

Run the script a couple times to verify that temperature is being accurately reported. Hold the external probe in your hand to warm it up, and watch that warmth reflect in the readings being displayed in the script's output.

Excessively Warm

Now that we're able to detect the ambient room temperature, let's use the same approach we did in Chapter 4, Water Leak Notifier, on page 49, and place upper and lower temperature boundaries to report when the temperature exceeds a set threshold condition. We're also only going to poll the external temperature probe, so be sure it's connected and working. In this example, set the temperature threshold to 80 degrees Fahrenheit:

```
huefan/temperaturetest.py
import smbus
import time

DEVICE_BUS = 1
DEVICE_ADDR = 0x17

TEMP_REG = 0x01
LIGHT_REG_L = 0x02
LIGHT_REG_H = 0x03
STATUS_REG = 0x04
ON_BOARD_TEMP_REG = 0x05
ON_BOARD_HUMIDITY_REG = 0x06
ON_BOARD_SENSOR_ERROR = 0x07
BMP280_TEMP_REG = 0x08
BMP280_PRESSURE_REG_L = 0x09
BMP280_PRESSURE_REG_M = 0x0A
BMP280_PRESSURE_REG_H = 0x0B
BMP280_STATUS = 0x0C
MOTION_DETECT = 0x0D

bus = smbus.SMBus(DEVICE_BUS)

alert_trigger = False

while True:
    aReceiveBuf = []

    aReceiveBuf.append(0x00)

    for i in range(TEMP_REG,MOTION_DETECT + 1):
        aReceiveBuf.append(bus.read_byte_data(DEVICE_ADDR, i))

    if aReceiveBuf[STATUS_REG] & 0x01 :
        print('The external temperature sensor range was exceeded.')
    elif aReceiveBuf[STATUS_REG] & 0x02 :
        print('The external temperature sensor is disconnected.')
    else :
        external_celsius = aReceiveBuf[TEMP_REG]
        external_fahrenheit = (external_celsius * 9/5) + 32
        if external_fahrenheit > 79:
            if alert_trigger != True:
                print("Temperature is above 79 degrees.")
                alert_trigger = True
        else:
            if alert_trigger != False:
                print("Temperature is below 80 degrees.")
                alert_trigger = False

    time.sleep(1)
```

Run the script and note the temperature readings being printed out in your Terminal window:

```
$ python3 temperaturetest.py
```

Breathe on or use a hair dryer on a low-heat setting to blow warm air over the external temperature probe. Once the temperature reading meets or exceeds 80 degrees, that status will be printed in the Terminal. Remove the heat source and wait for the temperature to drop. When the air around the probe drops below 80 degrees, the message Temperature under 80 degrees. will display in the Terminal running the script.

Watch the output screen as you raise and lower the temperature around the external probe, then escape out of the running temperaturetest.py script.

We now have the ability to control power switches and sense the temperature. It's time to combine all those separate capabilities into a single executable script.

All Together Now

Combine the code we used to toggle the Hue smart plug on and off with the external temperature test script. Doing so will allow us to turn the fan on if the temperature exceeds 79 degrees, and turn it off if the temperature drops below 80 degrees.

We also need to reduce the sampling rate to every two minutes (120 seconds) instead of every second. This will provide ample duration for the smart plug to turn on and off the fan. Otherwise, we'll be constantly turning the fan on and off as the temperature fluctuates between a single threshold degree:

```
huefan/huefan.py
from phue import Bridge
import smbus
import time

huebridge = Bridge('YOUR_HUE_BRIDGE_IP_GOES_HERE')

DEVICE_BUS = 1
DEVICE_ADDR = 0x17

TEMP_REG = 0x01
LIGHT_REG_L = 0x02
LIGHT_REG_H = 0x03
STATUS_REG = 0x04
ON_BOARD_TEMP_REG = 0x05
ON_BOARD_HUMIDITY_REG = 0x06
ON_BOARD_SENSOR_ERROR = 0x07
BMP280_TEMP_REG = 0x08
BMP280_PRESSURE_REG_L = 0x09
BMP280_PRESSURE_REG_M = 0x0A
BMP280_PRESSURE_REG_H = 0x0B
BMP280_STATUS = 0x0C
HUMAN_DETECT = 0x0D
```

```python
bus = smbus.SMBus(DEVICE_BUS)

alert_trigger = False

while True:
    aReceiveBuf = []
    aReceiveBuf.append(0x00)
    for i in range(TEMP_REG,HUMAN_DETECT + 1):
        aReceiveBuf.append(bus.read_byte_data(DEVICE_ADDR, i))
    if aReceiveBuf[STATUS_REG] & 0x01 :
        print('The external temperature sensor range was exceeded.')
    elif aReceiveBuf[STATUS_REG] & 0x02 :
        print('The external temperature sensor is disconnected.')
    else :
        external_celsius = aReceiveBuf[TEMP_REG]
        external_fahrenheit = (external_celsius * 9/5) + 32
        if external_fahrenheit > 79:
            if alert_trigger != True:
                huebridge.set_light('Fan','on', True)
                alert_trigger = True
        else:
            if alert_trigger != False:
                huebridge.set_light('Fan','on', False)
                alert_trigger = False
    time.sleep(120)
```

Save and execute the script.

```
$ python3 huefan.py
```

Tweak the temperature and sleep time (that is, sampling rate) values to match your ideal room-cooling conditions. Change the sampling rate further if you discover that the temperature fluctuates repeatedly between 79 and 80 degrees. This way you're not constantly turning on and off the smart plug every two minutes. You can also create a nested conditional statement that extends the duration of the smart plug turning on or off based on the last temperature reading. You can continue to tinker as much as you prefer to attain fine granularity over the triggers responding to the temperature variants in your room.

If you intend to keep the script running at all times, create and enable a Supervisor configuration file for the project and save it in the /etc/supervisor/conf.d directory:

huefan/huefan.conf
```
[program:huefan]
directory = /home/YOURUSERNAME/projects/huefan
command = /usr/bin/nohup /usr/bin/python3 huefan.py &
user = YOURUSERNAME
environment=HOME="/home/YOURUSERNAME",USER="YOURUSERNAME"
```

```
autostart = true
autorestart = true
stdout_logfile = /var/log/supervisor/huefan.log
stderr_logfile = /var/log/supervisor/huefan.log
```

Restart the Supervisor daemon to activate the `huefan.py` supervised Python script.

```
$ sudo service supervisor restart
```

Nice job! You've been able to parlay your knowledge from Chapter 4, Water Leak Notifier, on page 49, into essentially creating your own thermostat. You now have enough experience and coding skills to connect to and read from the Pi's inputs. You can even extend this project further by including the email alert code we wrote for the water sensor project. This way, you'll be notified that your fan was turned on or off even if you're not in the same room.

Next Steps

Now that you have the foundation to control appliances like fans based on external inputs like ambient light and room temperature, expand upon that base project. Create schedules to trigger such sensor tests, or enable other Hue-enabled accessories like light strips and wall switches to trigger at certain times or environmental conditions in the home or workplace.

Before leaving the exciting world of sensor management, we'll explore one additional sensor-related situation in our next project. Using the light sensor on the SensorHub coupled with code fragments assembled from our previously completed projects, we'll construct the means to turn on a light when motion is detected and turn it off after a set duration.

CHAPTER 6

Hue Auto Light

The concept of turning lights on when motion is detected has been operational for decades. Originally designed for industrial and commercial building spaces, manufactured solutions are expensive and often proprietary installations. But as we've seen with other projects in this book, the Pi has redefined the cost of computing and, with it, the cost to implement automation solutions.

Professional lighting systems allow operators to set a timer for how long motion is no longer detected before turning lights off. These systems also allow for lighting schedules to turn on and off lights depending on the time of day. Only the most expensive and sophisticated commercial installations also take into account if additional lighting is required based on existing light sources illuminating a room. Thanks to the Pi-SensorHub combination, our implementation will cost a tenth of that of traditional proprietary lighting systems yet perform the same functionality.

For this project, we're going to borrow designs from our previous sensor detector projects. We'll also utilize the SensorHub's motion detector to trigger lighting events, with a twist; we'll want to account for motion and time variants so we're not turning the light on and off every second. Let's get started.

Setup

Here's what you need to build this project.

Hardware

- DockerPi SensorHub
- Philips Hue bridge
- Philips Hue smart bulb[1]

---

1. https://www.philips-hue.com/en-us/products/smart-lightbulbs

You can use either the standard white or color light Hue smart bulbs. Although a bit more expensive than the white bulb, I prefer the color light model so I can dial in the perfect color for the environment or mood I'm in. Here's a photo of the Hue color bulb model in my hand.

Software

- phue Python library
- Philips Hue smartphone app

If you've completed Chapter 5, Hue Fan, on page 61, then you already have the phue Python library installed. If not, you can install the phue library via the pip3 install command.

```
$ sudo pip3 install phue
```

Assuming you successfully completed the last Hue project involving the SensorHub and that hardware add-on is still attached to your Pi, then you're ready to proceed. If not, follow the setup in that project to attach the SensorHub and activate the I2C interface. Then register your Pi with the Hue bridge by following the instructions in the Hue Fan Bridge Registration section.

And There Was Light

Make sure you've already configured and labeled your Hue smart bulb that you want to use in this project with the Hue bridge via the Hue smartphone app. Confirm that you can turn that particular smart bulb on and off from the Hue smartphone app before proceeding.

Next, identify that particular smart bulb you want to automate via the listde-vices.py script we used in Chapter 5, Hue Fan, on page 61:

on page 61

hueautolight/listdevices.py
```
from phue import Bridge

huebridge = Bridge('YOUR_HUE_BRIDGE_IP_GOES_HERE')

lights = huebridge.lights

for light in lights:
    print(light.name)
```

Identify the label of the preferred smart bulb name in the generated list, and then confirm you can turn on and off that bulb programmatically with a smart-bulbtest.py Python script. In my example, I'll be turning on and off the smart bulb I installed in my kitchen light fixture. Be sure to change the Kitchen placeholder for the name of the Hue smart bulb you are testing on your network:

hueautolight/smartbulbtest.py
```
from phue import Bridge
import time

huebridge = Bridge('YOUR_HUE_BRIDGE_IP_GOES_HERE')

huebridge.set_light('Kitchen','on', True)
time.sleep(10)
huebridge.set_light('Kitchen','on', False)
```

Just like the Hue Fan project, you should see your specified smart bulb turn on and then turn off ten seconds later. Be sure this script works correctly with your Hue setup before continuing with this project.

See how quickly that went? We're nearly halfway done with completing the project's objective. For the other half, we need to call upon the SensorHub's motion detector to identify when motion takes place:

hueautolight/motionsensortest.py
```
import smbus
import time

DEVICE_BUS = 1
DEVICE_ADDR = 0x17

TEMP_REG = 0x01
LIGHT_REG_L = 0x02
LIGHT_REG_H = 0x03
STATUS_REG = 0x04
ON_BOARD_TEMP_REG = 0x05
ON_BOARD_HUMIDITY_REG = 0x06
ON_BOARD_SENSOR_ERROR = 0x07
BMP280_TEMP_REG = 0x08
BMP280_PRESSURE_REG_L = 0x09
```

```
BMP280_PRESSURE_REG_M = 0x0A
BMP280_PRESSURE_REG_H = 0x0B
BMP280_STATUS = 0x0C
MOTION_DETECT = 0x0D

bus = smbus.SMBus(DEVICE_BUS)

alert_trigger = False

while True:
    aReceiveBuf = []

    aReceiveBuf.append(0x00)

    for i in range(TEMP_REG,MOTION_DETECT + 1):
        aReceiveBuf.append(bus.read_byte_data(DEVICE_ADDR, i))

    if aReceiveBuf[MOTION_DETECT] == 1 :
        if alert_trigger != True:
            print("Movement was detected.")
            alert_trigger = True
    else:
        if alert_trigger != False:
            print("No movement was detected.")
            alert_trigger = False

    time.sleep(1)
```

Recall from the SensorHub overview in the Hue Fan project that when motion is detected, a blue LED along the middle-right side on the SensorHub will light up. The motion detection event will also report its status when running our motionsensortest.py script. Try running the script now:

$ python3 motionsensortest.py

Wave your hand over the SensorHub and when the blue LED motion indicator illuminates, you should see a Movement was detected. message print out in the Terminal window where the script is running. It will be followed by a No movement was detected. message if no additional movement was picked up by the sensor after a one-second delay.

Light Bright

We don't need to turn the light on and unnecessarily waste electricity if there's enough natural daylight in the room already. To test for that condition, we'll poll the SensorHub's onboard light sensor. The light sensor will help determine if additional room lighting is necessary. The onboard SensorHub light sensor detects the amount of light energy present in units of lux.[2]

---

2.   https://en.wikipedia.org/wiki/Lux

For this project, we'll turn on the Hue smart bulb when motion is detected as long as the amount of available light detected in the room is less than 25 lux, which is roughly equivalent to the ambient light diffused in a slightly darkened room. Let's write a quick light-sensor test script to detect the light illumination value and react accordingly:

hueautolight/lightsensortest.py

```python
import smbus
import time

DEVICE_BUS = 1
DEVICE_ADDR = 0x17

TEMP_REG = 0x01
LIGHT_REG_L = 0x02
LIGHT_REG_H = 0x03
STATUS_REG = 0x04
ON_BOARD_TEMP_REG = 0x05
ON_BOARD_HUMIDITY_REG = 0x06
ON_BOARD_SENSOR_ERROR = 0x07
BMP280_TEMP_REG = 0x08
BMP280_PRESSURE_REG_L = 0x09
BMP280_PRESSURE_REG_M = 0x0A
BMP280_PRESSURE_REG_H = 0x0B
BMP280_STATUS = 0x0C
MOTION_DETECT = 0x0D

bus = smbus.SMBus(DEVICE_BUS)

while True:
    aReceiveBuf = []

    aReceiveBuf.append(0x00)

    for i in range(TEMP_REG,MOTION_DETECT + 1):
        aReceiveBuf.append(bus.read_byte_data(DEVICE_ADDR, i))

    lux = (aReceiveBuf[LIGHT_REG_H] << 8 | aReceiveBuf[LIGHT_REG_L])

    if aReceiveBuf[STATUS_REG] & 0x04 :
        print('The onboard light sensor is overloaded.')
    elif aReceiveBuf[STATUS_REG] & 0x08 :
        print('The onboard light sensor is not responding.')
    else :
        print('The current onboard light sensor brightness = %d lux.' % lux)

    time.sleep(1)
```

Shine a light source, such as a smartphone camera light, at the SensorHub while running the script. You should see the lux value increase significantly and our script react accordingly. Then escape out of the script.

We've Got Movement

With the smart bulb and light and motion sensor programmatically confirmed working, we can combine these together to turn the light on in low-light situations whenever motion is detected.

We'll alter the Hue smart plug code from the previous Hue project and swap out the temperature code with our motion test code. But unlike the Hue Fan project, we shouldn't immediately turn off the light when motion is no longer detected; otherwise our light will just flicker on briefly and turn off seconds later.

We'll do so in our script by keeping the light on for five minutes. And we'll only turn on the smart bulb when there isn't already enough light illuminating the room:

```
hueautolight/hueautolight.py
from phue import Bridge
import smbus
import time

huebridge = Bridge('YOUR_HUE_BRIDGE_IP_GOES_HERE')

DEVICE_BUS = 1
DEVICE_ADDR = 0x17

TEMP_REG = 0x01
LIGHT_REG_L = 0x02
LIGHT_REG_H = 0x03
STATUS_REG = 0x04
ON_BOARD_TEMP_REG = 0x05
ON_BOARD_HUMIDITY_REG = 0x06
ON_BOARD_SENSOR_ERROR = 0x07
BMP280_TEMP_REG = 0x08
BMP280_PRESSURE_REG_L = 0x09
BMP280_PRESSURE_REG_M = 0x0A
BMP280_PRESSURE_REG_H = 0x0B
BMP280_STATUS = 0x0C
MOTION_DETECT = 0x0D

bus = smbus.SMBus(DEVICE_BUS)

alert_trigger = False

while True:
    aReceiveBuf = []
    aReceiveBuf.append(0x00)
    for i in range(TEMP_REG,MOTION_DETECT + 1):
        aReceiveBuf.append(bus.read_byte_data(DEVICE_ADDR, i))

    lux = (aReceiveBuf[LIGHT_REG_H] << 8 | aReceiveBuf[LIGHT_REG_L])

    if aReceiveBuf[MOTION_DETECT] == 1 :
```

```
        if alert_trigger != True:
            if lux < 25 :
                huebridge.set_light('Kitchen','on', True)
                alert_trigger = True
                time.sleep(300) # Leave the light on for 5 minutes
    else:
        if alert_trigger != False:
            huebridge.set_light('Kitchen','on', False)
            alert_trigger = False

    time.sleep(1)
```

Save your work, run the code, and test in various scenarios:

```
$ python hueautolight.py
```

Change the lighting conditions around the SensorHub to verify that the light sensor is doing its job of allowing the smart bulb to turn on only when there isn't enough light already in the room. You can artificially simulate this by placing a box on top of the Pi-SensorHub to darken its environment. Conversely, if you're already in a dark room, shine a light on the sensor to artificially elevate the luminosity detected.

Once you're satisfied with the results, customize the conditional lux and timeout values to suit your own environment and automate the auto-execution of the script by creating its own hueautolight.conf configuration file for Supervisor to consume.

If you haven't already done so, follow the Supervisor setup instructions in Chapter 4, Water Leak Notifier, on page 49, and then create a hueautolight.conf file in the /etc/supervisor/conf.d directory:

```
$ sudo nano /etc/supervisor/conf.d/hueautolight.conf
```

The contents of this file will closely resemble the one we used in our previous projects, only changing the paths to and names of the scripts being run:

```
hueautolight/hueautolight.conf
[program:hueautolight]
directory = /home/YOURUSERNAME/projects/hueautolight
command = /usr/bin/nohup /usr/bin/python3 hueautolight.py &
user = YOURUSERNAME
environment=HOME="/home/YOURUSERNAME",USER="YOURUSERNAME"
autostart = true
autorestart = true
stdout_logfile = /var/log/supervisor/hueautolight.log
stderr_logfile = /var/log/supervisor/hueautolight_err.log
```

Activate the script's supervision in one of two ways. You can restart the Supervisor service:

```
$ sudo service supervisor restart
```

Or you can simply reboot the Pi:

```
$ sudo reboot
```

Run your lighting tests again to make sure the script is executing in the background. Great job! You've just created a custom lighting solution that would have cost tens of thousands of dollars to implement in a commercial building setting. And you did it within a matter of minutes due to the knowledge gained, along with the source code created, in the previous Pi projects you completed.

Next Steps

This project was straightforward, especially if you worked through the preceding Hue project controlling the fan. Polling the SensorHub's motion sensor and turning on the Hue-managed light via a Python script may seem like a simple affair, but the hardware and network infrastructure required to make it work is what makes everything seem so easy.

If you successfully completed both the Hue Fan and this Hue lighting project, congratulations! You've successfully mastered the ability to program your Hue system to do anything autonomously. You normally would need the Hue app to control lighting and electrical appliances manually, but now you have the power to make your Hue-powered systems perform multiple tasks based on sensor data and event changes.

Consider other areas of your home or office you'd like to automate. Turn on a coffeemaker at a certain time and monitor the temperature. Send an email reminder if your coffee has been sitting idle at the same temperature for too long. Automatically water plants based on the humidity of the soil. Roll up or down window shades based on the room temperature or time of day. All these ideas and more are possible now that you know how easy it is to implement them.

Our next project will take a departure from sensors and switches to focus on giving our Pi something to do in the middle of the night. Due to the project's complexity, we'll also need to write a bit more Python code than we've used in the book thus far. For now, you can keep cool with the Hue Fan and see easier with your Hue Auto Light.

CHAPTER 7

PiSpeak

The Pi 4 is remarkably powerful hardware, considering its diminutive size and relatively low cost. And yet, if you were to only use that computing power for monitoring occasional changes in sensors, a lot of compute cycles would be wasted. The purpose of this project is to give your Pi something to do during the off-hours that will hopefully make you more productive and even smarter when you're not sleeping.

The objective of the PiSpeak project is to retrieve the latest news using RSS feeds and use a text-to-speech (TTS) engine to convert those feeds into a single MP3 file that can be easily synced with a cloud storage service. Then that MP3 file can be downloaded to a smartphone and configured for playback in a smartphone MP3-player application. The script will also send an email notification to alert when new content is available along with a list of titles and links to those articles contained in the audio file for later reference.

This project features one of the longer scripts in the book. You can use whichever text editor you prefer, but I recommend using Microsoft Visual Code to help highlight the source code syntax, navigate, and debug your Python code. You can also use a Mac or Windows PC to write and test the script before ultimately deploying it to the Pi. Thanks to Python's portability, it's easy to code on one platform and run the script on a completely different OS and CPU architecture. Of course, if you prefer to work entirely on the Pi, that's fine too, since Visual Studio Code works identically on the Pi, albeit slower compared to more powerful Mac- or PC-hardware platforms. So put on your Python coder cap and let's get started.

Setup

Here's what you need to build this project.

Hardware

- Smartphone
- Speaker or headset to test audio playback
- (Optional) Mac or Windows PC for remote script development

Software

- beautifulsoup4[1] Python library
- feedparser[2] Python library
- ffmpeg[3]
- mutagen[4] Python library
- Path (part of the standard Python library)
- Rclone (already installed in Chapter 2, Setting Up the Software, on page 15)
- shutil (part of the standard Python library)
- (Optional) DB Browser for SQLite[5]
- (Optional) SQLite Extension for VS Code[6]
- MP3 Audiobook Player Pro for iOS,[7] or
- Smart Book Player for Android[8]

Before we begin installing the list of software dependencies, let's specifically break down what we want to achieve.

1. Retrieve a list of recent articles posted to a selection of favorite news sources.

2. Aggregate these articles into a single body of text that can be processed for TTS.

3. TTS process the text and save the output as an MP3 file.

4. Accelerate the audio playback without altering the pitch, and convert the output to a new MP3 file.

5. Tag this new, sped-up MP3 file with title, artist, and cover art metadata for more meaningful rendering in audio playback applications.

---

1. https://pypi.org/project/beautifulsoup4/
2. https://pypi.org/project/feedparser/
3. https://ffmpeg.org/
4. https://pypi.org/project/mutagen/
5. https://sqlitebrowser.org/
6. https://marketplace.visualstudio.com/items?itemName=alexcvzz.vscode-sqlite
7. https://apps.apple.com/us/app/mp3-audiobook-player-pro/id889580711
8. https://play.google.com/store/apps/details?id=ak.alizandro.smartaudiobookplayer

6. Move this final MP3 file into a folder that synchronizes with a cloud storage service of choice, making it simple to retrieve the file from a smartphone or media playback system that may or may not be on our home or office network.

7. Send an email to notify you what new content is available for retrieval and the name of the timestamped MP3 file that contains it.

8. Download the MP3 file from your cloud storage provider via smartphone or computer.

9. Queue up, playback, and listen.

Implementing this sequential list of tasks is straightforward, especially since the apps and libraries to help us do this have already been built by dedicated members of the programming community. Let's begin with retrieving a news article listing using Really Simple Syndication (RSS).

Feeding RSS

Back when RSS feeds were new, media companies were prominently promoting them as a means to subscribe and retrieve content updates. Over time, different means of notification and content consumption moved RSS into the background. There was also the potential confusion that nontechnical users had when clicking an RSS link only to see a bunch of XML code render in their browser window. Consequently, major media providers removed RSS references from their pages, but most news outlets still offer many of these RSS feeds if you know where to look for them.

For example, most blogging websites built on WordPress offer the ability to retreive their RSS feed by simply adding /feed/ after the site's domain name, such as the one for hackaday.com.[9] Depending on the web browser you're using, visiting that URL will render an XML document containing a number of various XML tags. Manually parsing and reading through such a document would be an eye-straining waste of time. Fortunately, Python can make these types of RSS document feeds intended for machines rather than humans to read much easier to consume and convert for our needs.

Feeding the Details

While you could use Python's built-in library to retrieve and parse RSS feeds, a popular third-party library called feedparser already does that for us. Install it via the usual pip3 install command:

---

9. https://hackaday.com/feed/

```
$ sudo pip3 install feedparser
```

You'll also need two other Python libraries. These will be used to remove any HTML tags, such as paragraph (<p>) or image (<image>) references. If we don't, these tags will be included in the text-to-speech conversion and make it unbearable to listen to (unless of course you enjoy hearing a series of angle brackets and tag names read back to you interspersed within the article text). A very popular tag parsing tool called Beautiful Soup 4, coupled with an easy-to-use Python HTML parser called lxml, will make the removal of these unwanted HTML tags as simple as one line of code. Install these libraries using pip3:

```
$ sudo pip3 install beautifulsoup4
```

```
$ sudo pip3 install lxml
```

You'll find a vast sea of RSS feeds to choose from, but to focus on a demonstrative few, in this project we'll consume them from Pragmatic Bookshelf–sponsored developer conversation website Devtalk.com,[10] National Public Radio News,[11] and the Pi enthusiast website, Raspberry Projects.[12]

The following Python script will iterate through each of these feeds, retrieving and displaying the recently posted article titles, reference web page links, and descriptive content. Using your text editor of choice, create a file called feedtest.py and enter the following code:

pispeak/feedtest.py
```python
❶ from bs4 import BeautifulSoup
  import feedparser

❷ URLS = ['https://forum.devtalk.com/latest.rss',
          'https://feeds.npr.org/1001/rss.xml',
          'https://projects-raspberry.com/news-updates/raspberry-pi-news/feed/',
          ]

❸ for url in URLS:
      feed = feedparser.parse(url)

❹     for entry in feed["entries"]:
          title = entry.get("title")
          link = entry.get("link")
❺         description = BeautifulSoup(entry.get("description"), "lxml").text
❻         print(title + '\n' + link + '\n' + description + '\n\n')
```

Let's quickly review the code used in this script before we attempt to run it.

---

10. https://forum.devtalk.com/latest.rss
11. https://feeds.npr.org/1001/rss.xml
12. https://projects-raspberry.com/news-updates/raspberry-pi-news/feed/

❶ Import the BeautifulSoup and feedparser Python libraries.

❷ Store the three RSS feed URLs that will be parsed in a static URLS array.

❸ Assign a single url from each URL stored in the URLS array, and process that RSS feed url by having feedparser parse the feed elements that can be later queried upon.

❹ For each entry in the feed being parsed, extract the title, link and description elements.

❺ Rather than just store the raw HTML description, remove the embedded HTML tags so they're not rendered when passing this string to the text-to-speech parser later in this project. Doing so will also make the output more human-readable as well.

❻ Combine the results of the title, link, and HTML-cleaned description into a string with appropriate line breaks, and print() out the results.

Save and run the feedtest.py file and execute it with the Python interpreter:

```
$ python3 feedtest.py
```

Assuming your code syntax is correct and your Pi is connected to the Internet, you should see a list of article titles and associated web links.

You now have a script that will quickly print out the latest newsfeed articles that have been posted to websites selected for this demonstration. That's pretty nifty. You also completed the project's first objective.

Next, you need to identify which of those listed articles were added since the last time the feedresults.py script was executed. Otherwise, you would have to listen to news items you previously heard sprinkled with any additions made to the feed since the last time the script was run. To accomplish that, retrieved key identifiers, namely the title and link entities, will be stored in a database. That database can be used to quickly check whether or not articles being retrieved have already been seen in previous feedresults.py results.

Database CRUD

While we could use a basic approach of storing the list of article titles and their unique web link identifiers to a plain-text file, we would then need to be responsible for the code to parse and update that file. As the file grew larger with more entries, the parsing would become slower due to more data to iterate through. And if we need to delete any data from the list, we would also need to maintain the integrity of the rest of the list via a lot of error-checking code.

Fortunately for our project, Python includes built-in support for a rather basic yet powerful and extremely popular database engine called SQLite.[13] While not nearly as feature-packed as other open source database engines like PostgreSQL,[14] SQLite is ideal for projects like ours that just need a local data store to perform create, read, update, and delete (better known as CRUD) operations. SQLite is lightweight and fast, and it can be called upon in Python with just a few simple lines of code.

The SQLite engine is already baked into Python, but it will be useful for us to view the data stored in the database we create from our Python script. A number of ways to accomplish this exist, with the most simple being to install the sqlite3 engine for Raspberry Pi OS from the Terminal:

```
$ sudo apt install sqlite3
```

Once installed, you can call upon this engine by entering sqlite3 at the Terminal's prompt. This will launch SQLite and present you with its own prompt. Learning the various commands SQLite has to offer is beyond the scope of this book, but I strongly recommend learning more about SQLite from websites such as SQL Tutorial.[15]

If you're running the Raspberry Pi OS Desktop edition, a helpful GUI-based SQLite database client is DB Browser for SQLite. You can install this using the apt command from the Terminal.

```
$ sudo apt install sqlitebrowser
```

I'll be predominantly using this application to display screenshots of this project's queries, due to its ease of use and clean tabular presentation of data.

Lastly, if you're using Microsoft Visual Code as your editor of choice, you can install a popular SQLite extension called, well, SQLite.[16] While not as full-featured as the DB Browser app, it's functional enough for viewing query results and performing basic SQLite database operations.

With your choice of SQLite viewer installed, we can proceed with adding database creation and storage capability to the code we previously wrote. Create a new Python script file called feeddbtest.py, copy and paste the feedtest.py code, and then add the extra SQLite code to it:

---

13. https://sqlite.org/
14. https://www.postgresql.org/
15. https://www.sqlitetutorial.net/
16. https://marketplace.visualstudio.com/items?itemName=alexcvzz.vscode-sqlite

pispeak/feeddbtest.py

```
from bs4 import BeautifulSoup
import feedparser
import os
① import sqlite3

   URLS = ['https://forum.devtalk.com/latest.rss',
           'https://feeds.npr.org/1001/rss.xml',
           'https://projects-raspberry.com/news-updates/raspberry-pi-news/feed/',
           ]

② if os.path.exists('feeddata.db'):
       read_articles = True
   else:
       read_articles = False
       print("feeddata.db database doesn't exist, creating new database file.")
③     conn = sqlite3.connect('feeddata.db')
       c = conn.cursor()

       c.execute('''CREATE TABLE NewsFeeds (
           id INTEGER PRIMARY KEY AUTOINCREMENT,
           title TEXT,
           link TEXT
④         );''')

⑤     conn.commit()
       conn.close()

   conn = sqlite3.connect('feeddata.db')
   c = conn.cursor()

   for url in URLS:
       feed = feedparser.parse(url)

       for entry in feed["entries"]:
           title = entry.get("title")
           link = entry.get("link")
           description = BeautifulSoup(entry.get("description"), "lxml").text

⑥         c.execute("SELECT link FROM NewsFeeds WHERE link=?", [link])
           link_exists = c.fetchone()
           if link_exists:
               # Record already extists in the database
               pass
           else:
               newdata = (title, link)
               c.execute("INSERT INTO NewsFeeds (title, link) VALUES (?,?)",\
⑦             newdata)
               print("Added to DB: " + title)

   conn.commit()
   conn.close()
```

Here's a quick review of the new code we added to the script.

❶ Import the sqlite Python library.

❷ Check to see if the feeddata.db database file already exists. If this is the first time we run the script, we'll need to create it.

❸ Connect to the feeddata.db database file. Python will automatically create an empty feeddata.db file for us if one doesn't already exist.

❹ Create a new table called NewsFeeds in our feeddata.db database with fields for id, title, and link.

❺ commit the database table changes to the feeddata.db database file and close the connection to make those changes permanent.

❻ Execute a SELECT statement to see if we already stored the record in the database before. If we did, there's no reason to store it again.

❼ If this is in fact a new record that we haven't seen before, we'll add it to the database using the INSERT statement. Then we commit any changes made and close the connection to the feeddata.db database.

Save the feeddbtest.py file and run it:

```
$ python3 feeddbtest.py
```

Since this is the first time you've run the script, it will create a new feeddata.db database file and then populate the NewsFeeds table with the most recent title and link entities retrieved from the RSS feeds. If you try running the script again after the script's initial successful run, the script will see that the feed-data.db file already exists, so it won't re-create or overwrite the NewsFeeds table and the data stored within it. And unless there was a newsfeed that updated its content in between your script executions, there will likely be no new data added to the database either.

To see what data was stored in the NewsFeeds table, launch the DB Browser for SQLite app (located in the Programming submenu on the Raspberry Pi OS Desktop), then navigate to and open the feeddata.db database file. Select the Browse Data tab and scroll through the data you stored there.

Nice job! You've just completed the most challenging portion of the project. Now that you know what new articles are available, you just need to convert them to MP3 audio format, then store that compiled audio file into the previously designated Rclone'd folder for easy cloud-based file retrieval.

Converting Audio

Several approaches are available to convert text to speech, but the easiest I've come across is an incredibly useful and powerful Python-based library called gtts. You can install it using the pip3 command:

```
$ sudo pip3 install gtts
```

Once installed, you can call its command-line interpreter front end in the Terminal, like any other program.

For example, if you wanted to generate a sample TTS file from this gtts-cli utility, you can do so from the Terminal window:

```
$ gtts-cli "This is a quick test." -o output.mp3
```

Locate the output.mp3 file in the same directory you ran the command, and using the File Manager desktop app, double-click the file to load and play it in the VLC media player.

Now let's do the same thing programmatically by calling upon the gtts library from a Python script. Using your preferred text editor, create a new file called ttstest.py and enter the following code:

```
pispeak/ttstest.py
❶ from gtts import gTTS

test_text = 'This is a test of TTS conversion for the PiSpeak project!'
❷ output = gTTS(text=test_text, lang='en', tld='com', slow=False)
❸ output.save("output.mp3")
```

Here's a brief explanation of what this script does.

- ❶ Import the gTTS library used to perform the TTS conversion.

- ❷ Pass the test_text text string along with language and playback speed parameters into the gTTS object, perform the conversion, and save the results to the output object.

- ❸ Save the audio content stored in the output object to the output.mp3 audio file.

Save and execute the ttstest.py script. Using the File Browser, open the output.mp3 file located in the same directory from where you ran the script and listen to the playback.

One annoying limitation with the current gtts implementation is its inability to selectively alter the TTS playback speed. I find even after setting gtts's slow parameter equal to False that the playback result is still too slow for my fast-paced listening preference. Fortunately there's an easy fix to this problem.

Processing the Audio

ffmpeg is one of the most important and popular open source utilities for rendering and manipulating audiovisual files. It is what powers the VLC player and numerous other open source and commercial media applications. In addition to being the Swiss Army knife of the media file world, it's also a tremendously useful tool for converting media files from one format to another. It can also be used to process additional effects on media data such as fades, transitions, and playback speed.

I'm always busy learning new things, with time being a key parameter in life that I can't change. But I can change how I use that time, as well as how I can use utilities like ffmpeg to change time for the content I consume. If you've ever used a modern podcast application, you know that most have the ability to speed up playback without affecting the pitch. In other words, you can listen to the same content in less time without the presenters sounding like chipmunks.

While the audio playback apps I listed in this project's Software section possess the built-in ability to speed up playback, it does take additional computational overhead on the smartphone to do so on the fly. If you plan on setting the playback speed on the device consistently at the same altered rate, then you're downloading twice as much data as you need to render the file. To reduce this overhead and take advantage of the available processing time we have on an idle Pi, we can have ffmpeg perform this file processing ahead of time. We'll do this for the MP3 file we previously generated.

ffmpeg is already installed in the Raspberry Pi OS for Desktop, but if you're using a different version of the OS, you can install ffmpeg via the apt install command:

```
$ sudo apt install ffmpeg
```

To further modify the playback speed of the rendered output.mp3 audio file, it will be passed into the ffmpeg program, along with an instructional filter:a "atempo" flag. The atempo filter will change the tempo or playback speed of the file. When atempo is set to 1.0, it won't alter the playback speed. To change the playback speed to twice as fast, set this value to 2.0.

Call upon ffmpeg to double-speed the playback and save the altered file to faster.mp3. We're going to instruct ffmpeg to apply this effect using the -vn parameter. We're also going to add the -y parameter to the command-line statements. Doing so gives ffmpeg permission to overwrite existing files with the same name. This way, you don't have to manually type Yes each time you run this conversion to listen to how the playback was altered. The -y command-line switch will also be useful in our final script for this project, since the script will keep running if an audio file by the same name already exists. If it does, we're allowing ffmpeg to simply overwrite the file. Here's the full command-line syntax to pass to the ffmpeg program:

```
$ ffmpeg -i output.mp3 -filter:a "atempo=1.5" -vn -y faster.mp3
```

Open the converted faster.mp3 file and listen to how much quicker the audio plays back compared to the original output.mp3 file. Also, take a look at the file size of the original output.mp3 compared with the converted faster.mp3. It should be roughly 25 percent of the original size. That will come in especially handy when syncing larger files with our devices. Not only will we use less bandwidth, but the transfer time will also be faster.

Continue to play around with the atempo values to find a speed that best suits your listening preference. Be aware that the atempo filter has a range between 0.5 and 2.0. If you want to speed up the playback even faster, you'll need to add the atempo filter twice. For example, try this if you want to quadruple the playback speed:

```
$ ffmpeg -i output.mp3 -filter:a "atempo=2.0,atempo=2.0" -vn -y faster.mp3
```

If you're a super-listener, you may be able to comprehend content played back at that speed. It's unintelligible to my ears, so I'll stick with my personal sweet spot of atempo=1.5 for use in the final script.

Tagging the Audio

Now that we have a converted, sped-up faster.mp3 file, we could simply leave the new file as is and finish up the project. But while the faster.mp3 file will play back just fine on the audio apps we'll listen to it with, the presentation won't be all that helpful. The faster.mp3 file is missing important metadata like the title and the author of the content. It also doesn't have any cover art, so our audio player programs will substitute whatever placeholder art for the missing cover art we failed to embed in the file.

We can easily address these deficiencies by calling upon the Python mutagen third-party library. Mutagen is a library that can be called upon to manipulate a variety of metadata stored in an MP3 file. Depending on how comprehensive the creator of an MP3 file wants to be, MP3 metadata can include the title, artist, album, year, genre, comments, and cover art among other things. For our project needs, we'll focus mainly on the title, artist, and cover image.

To keep things simple and consistent, we'll use a single JPEG file as the same cover image for our converted MP3 files. For readers who downloaded the code for this book, I've saved a smaller JPEG of this book's cover, saved as home.jpg, as the image I'll use for these examples.

Install the Mutagen library via the usual pip3 install command.

```
sudo pip3 install mutagen
```

Here's a simple test script on the faster.mp3 file to embed the title, artist/author, and home.jpg image into the MP3 file:

```
pispeak/mp3tagging.py
```
❶ ```
from mutagen.mp3 import MP3
from mutagen.easyid3 import EasyID3
from mutagen.id3 import ID3, APIC, error
```
❷ ```
mp3_file = EasyID3('faster.mp3')
```
❸ ```
mp3_file["title"] = "Sample Title"
mp3_file["artist"] = "Content Creator"
mp3_file["album"] = "PiSpeak Project"
mp3_file.save()
```
❹ ```
mp3_file = MP3('faster.mp3', ID3=ID3)
mp3_file.tags.add(APIC(mime='image/jpeg',type=3,desc=u'PiSpeak Project',
                  data=open('home.jpg','rb').read()))
mp3_file.save()
```

Let's take a look at what this script does.

❶ Import specific MP3 tagging functions from the mutagen Python library.

❷ Assign the mp3_file variable to the faster.mp3 file using the EasyID3() function. This function allows us to easily manipulate the MP3 tags without a lot of fuss.

❸ Embed the MP3 title, artist, and album metadata into the test.mp3 file. We've assigned placeholder text to these fields in this sample. In the final script, they'll change to match the respective entities retrieved from the download.

❹ Embed the home.jpg image as the cover art for the test.mp3 file. This art will be displayed in applications that parse and render embedded cover art in MP3 files, such as File Manager and certain media collection and playback apps.

Save the script as mp3tagging.py and run it, making sure to have the home.jpg and faster.mp3 source files in the same project directory as the script:

```
$ python3 mp3tagging.py
```

Take a look at the updated faster.mp3 file in the File Manager. You should now see it rendered with the book's cover. Try running the script a few more times, changing the title and artist and using a different JPEG file as your home.jpg cover to embed in the MP3 file.

We're getting close to completing all of our project's objectives. We have one more step to go before we can assemble these programming snippets into a single, complete Python script.

Move It

The last objective is to programmatically move the faster.mp3 file into the GoogleDrive folder that was created when Rclone was configured in Rclone, on page 28. To do so, call upon Python's built-in shutil library to move and rename the file. Create a movefile.py Python script in the same directory as the faster.mp3 file and use the following code to move the file into the GoogleDrive folder:

```
pispeak/movemp3file.py
from pathlib import Path
import platform
import shutil

homedirectory = str(Path.home())

if platform.system() == 'Windows':
    syncfolder = '\\GoogleDrive\\'
else:
    syncfolder = '/GoogleDrive/'

mp3name = "faster.mp3"

mp3destination = homedirectory + syncfolder + mp3name

shutil.move(mp3name, mp3destination)
```

Let's do a quick review of the listing before we run the script.

❶ Import the necessary pathlib, platform and shutil Python libraries used to identify the current working and home directory paths and operating system that the script is running on.

❷ Convert the Path.home() object into a string to obtain the file path to the current user's home directory.

❸ Call the platform.system() function to see what operating system is running the script. If it's Windows, change the directory separators to back slashes.

Otherwise, in the case of Mac and Linux, make them forward slashes to match the Unix file path specification.

❹ Assign the mp3name variable to the name of the file being moved; in this case, it's the faster.mp3 file.

❺ Concatenate the homedirectory + syncfolder and mp3name string assignments into the full file path and assign it to the mp3destination variable.

❻ Call the shutil.move() function, passing in the mp3name and mp3destination to move the MP3 file to the GoogleDrive directory in the user's home directory.

Save the file and run the movemp3file.py script":

```
$ python3 movemp3file.py
```

You just finished the final piece of the project. Now it's time to assemble these individual code snippets into a single, cohesive script that'll perform these steps sequentially.

In the final project script, you'll add a mail notification. The body of this email will list a summary of new article titles and links since the last time the script was run. It will also indicate the name of the new MP3 file that's been compiled and transferred to Google Drive for download. Lastly, the script will include instructions to delete the original output.mp3 source file. This will help continue to keep the script's working directory clean. When you're ready, proceed to the next section listing the final, complete script for the project.

List It

Here's the full script listing for the project. If you've been following along and running the test scripts up to this point, the elements in this listing should be familiar to you. We also lifted the send_email() function from our prior projects to notify via email when the script has completed all its tasks successfully.

To keep the script clutter-free so you can concentrate on what each line of code is doing, there won't be any numbered bullet call-outs. We've already discussed most of what the script does in the test scripts we already ran, so I'll only touch upon a few items in the summary after the listing. These are mostly the sections I've commented using the # symbol within the code listing itself.

If you downloaded the code for this book from the Pragmatic Bookshelf website, you should be able to execute the pispeak.py Python script directly within the code\pispeak directory (assuming that you already installed the required third-party libraries and have created a folder called GoogleDrive in your computer or Pi's home directory).

```
pispeak/pispeak.py
import feedparser
from bs4 import BeautifulSoup
from gtts import gTTS
from mutagen.mp3 import MP3
from mutagen.easyid3 import EasyID3
from mutagen.id3 import ID3, APIC, error
import os
from pathlib import Path
from pathlib import Path
import platform
import sqlite3
import shutil
import time
import smtplib
import email.mime.multipart
from email.mime.text import MIMEText

URLS = ['https://forum.devtalk.com/latest.rss',
        'https://feeds.npr.org/1001/rss.xml',
        'https://projects-raspberry.com/news-updates/raspberry-pi-news/feed/',
        ]

def send_email(message):
    gmail_user = 'YOUR-GMAIL-ACCOUNT-NAME@gmail.com'
    gmail_password = 'YOUR-GMAIL-ACCOUNT-PASSWORD'

    try:
        msg = email.mime.multipart.MIMEMultipart()
        msg['to'] = 'Your Name<YOUR-EMAIL-ADDRESS@DOMAIN.COM>'
        msg['from'] = 'Your Gmail<YOUR-GMAIL-ACCOUNT-NAME@gmail.com>'
        msg['subject'] = 'New PiSpeak Content Available'
        msg.add_header('reply-to', 'YOUR-GMAIL-ACCOUNT-NAME@gmail.com')
        msg.attach(MIMEText(message, 'plain'))
        session = smtplib.SMTP('smtp.gmail.com', 587)
        session.starttls()
        session.login(gmail_user, gmail_password)
        message = msg.as_string()
        session.sendmail('YOUR-GMAIL-ACCOUNT-NAME@gmail.com',
        'YOUR-EMAIL-ADDRESS@DOMAIN.COM', message)
        session.quit()
        print('Email sent.')
    except:
        print('Email failed to send.')

if os.path.exists('feeddata.db'):
    read_articles = True
else:
    read_articles = False
    print("feeddata.db database doesn't exist, creating new database file.")
    conn = sqlite3.connect('feeddata.db')
    c = conn.cursor()
```

```python
    c.execute('''CREATE TABLE NewsFeeds (
        id INTEGER PRIMARY KEY AUTOINCREMENT,
        title TEXT,
        link TEXT
        );''')

    conn.commit()
    conn.close()
conn = sqlite3.connect('feeddata.db')
c = conn.cursor()

storycount = 0
storycollection = 'The following new articles have been compiled for you:\n\n'
tobeconverted = ''

for url in URLS:
    feed = feedparser.parse(url)

    for entry in feed["entries"]:
        title = entry.get("title")
        link = entry.get("link")
        description = BeautifulSoup(entry.get("description"), "lxml").text

        c.execute("SELECT link FROM NewsFeeds WHERE link=?", [link])
        link_exists = c.fetchone()
        if link_exists:
            # Record already exists in the database
            pass
        else:
            newdata = (title, link)
            c.execute("INSERT INTO NewsFeeds (title, link) VALUES (?,?)",\
             newdata)
            storycount = storycount + 1
            storycollection = storycollection + title + '\n' + link + '\n\n'
            tobeconverted = tobeconverted + '\n' + title + '\n' + description

if tobeconverted != '':
    tobeconverted = tobeconverted + ' End of playback.'
    output = gTTS(text=tobeconverted, lang='en', tld='com', slow=False)
    output.save('output.mp3')

    os.system('ffmpeg -i output.mp3 -filter:a "atempo=2.0" -vn -y faster.mp3')
    os.remove('output.mp3')

    mp3_file = EasyID3('faster.mp3')

    mp3_file["title"] = "Sample Title"
    mp3_file["artist"] = "Content Creator"
    mp3_file["album"] = "PiSpeak Project"
    mp3_file.save()

    mp3_file = MP3('faster.mp3', ID3=ID3)
    mp3_file.tags.add(APIC(mime='image/jpeg',type=3,desc=u'PiSpeak Project',
                    data=open('home.jpg','rb').read()))
    mp3_file.save()
```

```
    # Add the time stamp and article count in final MP3 filename
    current_seconds = time.time()
    current_time = time.localtime(current_seconds)
    current_directory = os.getcwd()

    newsMP3name = 'NEWS - %d-%d-%d at %d-%d - %s articles.mp3' % (
    current_time.tm_year, current_time.tm_mon, current_time.tm_mday,
    current_time.tm_hour, current_time.tm_min, storycount)

    # Rename faster.mp3 to the one with more details in the final filename
    os.rename('faster.mp3', newsMP3name)

    # Move the final MP3 file to the GoogleDrive directory
    homedirectory = str(Path.home())
    if platform.system() == 'Windows':
        syncfolder = '\\GoogleDrive\\'
    else:
        syncfolder = '/GoogleDrive/'
    mp3destination = homedirectory + syncfolder + newsMP3name
    shutil.move(newsMP3name, mp3destination)

    message_body = storycollection + '\n\nready to be retrieved in '\
                                    + newsMP3name

    send_email(message_body)
conn.commit()
conn.close()
```

We've already individually reviewed the most interesting portions of this script in the sections leading up to this code listing. A few new additions are worth mentioning. For example, adding both title and description to the tobeconverted string aids with audio queues during playback that mark the the start and end of a new article. Additionally, tacking on the string End of playback. to the end of the tobeconverted string signals to the listener that audio playback is done. Finally, note that the built-in Python time library was used to timestamp the final MP3 filename, making it easy to identify when the file was compiled and how many articles are contained in its playback.

Run the script just like all other Python scripts:

```
$ python3 pispeak.py
```

The first time the script is run, it won't generate any audio content, because the database doesn't yet exist. Once the database is created and populated with new content since the last time you ran the script, audio processing will commence. Test this by randomly deleting a few of the records stored in the feeddata.db database. Using DB Browser on the Pi, open the feeddata.db database, choose the Browse Data tab and randomly click on any row of data, and select the Delete Record button in the upper-right corner of the Browse Data window pane. Then rerun the script. When the script successfully finishes this second time, check your email inbox to see what new content has been captured and converted for you into the timestamped MP3 file that was generated and saved to your Google Drive.

When you're writing a more involved Python script like this, it's best to break it down into smaller portions like we did throughout the chapter. Not only will you better understand and tackle each data processing stage with clarity and assurance that the routine works but you'll also learn more Python practices along the way.

Even after all the years I've been coding in Python, I still learn a few new tricks from Python posts on sites like Stack Overflow and various Python podcasts. When you are writing your own Python scripts, call upon these resources to give you ideas or quick solutions to your coding conundrums.

Play It On the Go

With the NEWS - ... .mp3 file moved into the previously configured Rclone GoogleDrive folder, the file is now accessible from your Gmail/Google Drive account. You can download this file from Google's cloud using a web browser or the native Google Drive app for your platform of choice. Once the Google Drive app is installed on iOS, you can use Apple's built-in Files app to access your Google Drive files, as shown in the next screenshot from my iPhone.

The iOS Files app is the file browser of choice for iOS applications. It works perfectly with Oleg Brailean's MP3 Audiobook Player Pro. This premium app is my preferred audiobook player on iOS, mainly due to its full-featured flexibility of consuming content from a number of sources. For example, accessing the NEWS - ... .mp3 file is as simple as accessing the app's ADD screen and selecting the Import via Cloud services option. This displays the Apple Files overlay where you can navigate to the Google Drive share and select the NEWS - ... .mp3 file uploaded therein.

MP3 Audiobook Player will then unzip and register the contents as a new "News Playlist" book, complete with the cover art that was embedded in the MP3 files it queued.

If you use Android as your preferred mobile device, Alex Kravchenko's Smart AudioBook Player is a free audiobook player with a worthy in-app purchase option needed to unlock its more comprehensive features. Unlike MP3 Audiobook Player Pro, Smart AudioBook Player doesn't include the ability to download content directly from cloud storage. Instead, you'll need to rely on the Google Drive Android application to download and store the file in the designated Smart AudioBook directory on your Android device. But that one additional step on Android is relatively insignificant, especially since I find Smart AudioBook Player's presentation more attractive than MP3 Audiobook Player Pro.

Select your cloud-stored NEWS MP3 file, queue it up, and press play on your preferred media player. Sit back, relax, and enjoy listening to the fruits of your labor. Once you've had your fill of audio, prepare for one last configuration to schedule on the Pi. Doing so will allow the script to automatically run at regular intervals so we have a collection of compiled content to listen to while we work or play.

Schedule It

Scheduling a Python script to run at set intervals is most frequently done on Unix-based operating systems like the Raspberry Pi OS as a command run on (UNIX scheduler) or CRON job, for short. Recall that we previously set up a cron job to mount the Rclone-configured GoogleDrive folder mount point at boot up.

By default, cron jobs are run by the system and not the user who added the job to the scheduler. Since we added all the Python library dependencies for this project under the pi user account, we need to make sure that the system runs the script as the pi user with the appropriate file paths to the libraries that pip installed. Create a new file in the project directory, called runpispeak.sh that will be the shell script that will establish these path and user configurations, and run the pispeak.py script.

```
pispeak/runpispeak.sh
#!/bin/sh
HOME=/home/pi
LOGNAME=pi
PATH=/usr/bin:/bin:/usr/local/bin/
SHELL=/bin/sh
PWD=/home/pi
cd /home/pi/projects/pispeak && python3 pispeak.py
```

The HOME and LOGNAME environment variables are set to the user's home directory and login name. Note that the PATH variable adds the /usr/local/bin/ file path to the standard system paths. When pip3 installs Python libaries at the sudo level, it stores these in the /usr/local/bin/ directory. The remaining SHELL and PWD variables are set to the standard shell and home working directory for the user. We conclude the shell script with the line that will actually change to the /home/pi/projects/pispeak project directory and execute the python3 pispeak.py script.

Save the runpispeak.sh file and exit the editor.

Before we can use the runpispeak.sh shell script, we need to tell the operating system that this is an executable file. We do that using the chmod +x command in the Terminal.

$ chmod +x runpispeak.sh

With the file marked as executable, test it out by running it from the Terminal.

$ ./runpispeak.sh

This should launch the pispeak.py script and run it as expected. Make sure this works correctly before proceeding.

With our working shell script ready to go, we can create a cron job to run the script every morning at 3:30 AM local time. This should give the Pi plenty of time to download, process, zip, move, and sync the audio collection to the cloud. To do so, run the crontab -e program in the Terminal:

$ crontab -e

Scroll to the bottom of the crontab file, below the previously added @reboot line for mounting GoogleDrive, and add the following cron command.

30 3 * * * /home/pi/projects/pispeak/pispeak.sh 2>&1

The first number in the crontab command is the minute mark, followed by the hour, day of the month, month, and day of week. Since we want this cronjob to run at 3:30 AM every day of every month of every year, we set the first number to 30 followed by 3. The * character is inclusive of all values, instructing the cron job to run the assigned command every date, month, and day of the week.

The main string is the command we want the system to run at that time. In this case, it is /home/pi/projects/pispeak/pispeak.sh 2>&1. The 2>&1 instruction at the end of the script command instructs cron to redirect 2, a file description for standard error (stderr), to 1, a file descriptor for standard output (stdout). This makes it easy to debug problems with a cronjob since it combines both output and errors into a single file. If we needed to perform additional debugging, we could rewrite the line to include the file path and name of the file to store this output.

/home/pi/projects/pispeak/pispeak.sh > /home/pi/projects/pispeak/output.txt 2>&1

I suggest creating this output.txt file only when debugging. Otherwise you'll have a file that will inflate each time the pispeak.sh shell script is run.

Save and exit the editor. The crontab should report that crontab is installing new crontab. You can either wait until 3:30 AM to see if the script executes on its own or re-edit the cron job and change the time to a few minutes after your current time so you can see it automatically execute the script sooner. Keep in mind that cron jobs use twenty-four-hour time, so setting the hour for, say, 7:00 PM would actually be set to 19. If you use 7 in the hour position instead, then the script will run at 7:00 AM.

With your Pi running the script every morning, you now have a continuous supply of entertaining and stimulating content to listen to as you clean the house, wash the car, cut the lawn, or just rest with eyes closed on the couch. And with the Pi doing the number crunching on speeding up playback, you can listen to twice as much in the same amount of time while easily multitasking other things. It's like having your own podcast network, and you are its sole customer.

Next Steps

Whew! That was a long project, but it was worth it. You now have a fully functional, audio-generating RSS reading application that can be tailored to whatever well-formed RSS feeds you want to parse. And because you're storing past feed results in a database, only new articles since the last time you ran the pispeak.py script will be compiled into a new MP3 file for listening at home, work, or on the go.

Improvements are guaranteed with any software endeavor, and this project is no different.

The TTS technology called upon by the gtts library is far better than the tinny robotic voices used by open source TTS engines like Festival,[17] but they still sound unnatural. Advanced machine learning is being used by major tech companies like Google, Microsoft, and Mozilla to generate high-quality TTS translations that sound much more like a speaking person. Using the sophisticated open source Mozilla TTS project[18] can make a huge difference in natural sounding text translation quality. Unfortunately, while installing and configuring Mozilla's TTS engine on a Pi 4 is possible, it is by no means as simple as running a pip3 install command. Read the Wiki on the Mozilla TTS project page for more details.

Another necessary improvement would be to wrap each phase of the workflow in try/except/finally blocks to handle exceptions. If an exception occurs, attempt to handle it programmatically or, at the very least, send an email indicating what kind of failure occurred.

For those who prefer a more stimulating challenge, construct a custom podcast client script that not only monitors what was downloaded and processed but what was also played back via the Pi's HDMI or analog headphone jack output. Mark what was listened to and either archive or delete the MP3 files that are no longer needed.

The next project will be a return to less Python code and more reliance on third-party services. These will assist with the ability to turn on computers anywhere in the world just by using the sound of your voice.

---

17. https://www.cstr.ed.ac.uk/projects/festival/
18. https://github.com/mozilla/TTS/

CHAPTER 8

Voice Wake on LAN

Smart voice assistants like Amazon Alexa or Google Assistant have become part of the home and office landscape, offering easy ways to initiate actions and respond to queries based on voice prompts. These services provide compelling generic access to prominent home-automation accessories like Philips Hue and certain high-end appliances.

However, responding to highly customized actions requires a great deal of application development expertise as well as experience working with the various APIs, coding, and configuration management dependencies that these assistants require. Even after such custom services are brought to life, their implementation is hampered by calling upon byzantine spoken word structures just to initiate a custom action. For example, to query Google Home on the delivery status of a UPS package, users must first log in to the Google Assistant, link their UPS account (assuming they even have one), and then ask to "Talk to UPS." The experience and even the voice changes with UPS's own Interactive Voice Response (IVR) menu options.

Fortunately, there's an easier way to ask your digital assistant to perform a simple custom task. Doing so requires the use of two additional service providers: IFTTT and Pushbullet. Both offer free-tier accounts. If you like their services and want to use more of them for your own projects beyond just those described in this book, you're welcome to consider subscribing to them.

We demonstrated in Chapter 7, PiSpeak, on page 83, how the Pi can be used for automatically processing workflows day or night, optimally at the times when the Pi is running idle. Doing so required kicking off the workflow at predefined intervals by means of setting up a cron job. But we can also leverage the Pi to respond to messages dynamically and act on these messages immediately, as opposed to a scheduled pull approach.

In this project, we're going to use our voice to ask our Google Home and/or Home Mini (or Amazon Alexa products for those who prefer Amazon smart assistants) to wake up and turn on a computer attached to a wired Ethernet connection on the same local area network segment as the Pi. This wake up service, known as Wake on LAN (WOL), is a popular way to remotely power up a Mac or PC hibernating in sleep mode. While some advanced wireless adapters support wireless WOL, my experience with them is that they're not as reliable as wired connections. For this project to work consistently, you should have the target Mac or PC connected via wired Ethernet on the same network as the Raspberry Pi that will be sending the WOL packets to it.

Let's get started!

Setup

Here's what you need to build this project.

Hardware

- Google Home, Home Mini, or Amazon Alexa Smart Assistant
- Mac or PC desktop using a wired Ethernet connection

Software

- wakeonlan[1]
- pushbullet[2] Python library

Services

- IFTTT[3]
- Pushbullet[4]

If you haven't already done so, associate your Google Home or Amazon Alexa smart speaker device with the Google or Amazon account of your choice. Voice request a few queries to make sure Google or Amazon responds in kind before proceeding.

Listen to Wake Up

Decide which desktop Mac or PC (connected to your wired local area network) that you want to remotely wake from sleep, as you will send it WOL network

---

1.   https://packages.debian.org/buster/wakeonlan
2.   https://github.com/rbrcsk/pushbullet.py
3.   https://ifttt.com/
4.   https://www.pushbullet.com/

instructions from the Pi. You will need to know the MAC address of that Mac or PC's Ethernet adapter port. Just like in the real world, you need to know the correct address of a recipient before you can send them a message.

To determine the MAC address of the Ethernet adapter on a Windows PC, open a Windows Command Prompt window (you can find the shortcut for the Command Prompt app in the Windows System folder on the Start menu), run the ipconfig program, and pass it the /all option:

```
> ipconfig /all
```

Doing so will list all the details of the network adapters on that PC. Look for the Ethernet adapter Ethernet entry with a Connection-specific DNS Suffix of lan. Assuming it's active with an assigned IP address that matches the default gateway used by your Pi, note the Physical Address associated with that adapter. Also note its Description, as you need that to identify the physical ethernet adapter in the Windows Device Manager.

If you're targeting macOS, open the Terminal app (located in the Mac's Applications -> Utilities folder) and run the ifconfig command. Look for the eth0 entry.

The pattern of the MAC address will be XX-XX-XX-XX-XX-XX, where the X placeholders will be replaced with the letters A through F and the numbers 0 through 9. The adapter address is unique to that physical adapter, and it's that address that we will pass to the wakeonlan program we'll install on the Pi. Take note of this MAC address before proceeding.

Configure your PC or Mac to listen for Wake on LAN packets. To do so on a PC, right-click the Windows Start menu icon and select Device Manager from the pop-up menu that appears. Expand the Network Adapters category and look for the name of the wired adapter you saw in the Description of the adapter in the ipconfig /all results. Right-click that specific network adapter in the Device Manager and select Properties from the pop-up menu. Select the Power Management tab and make sure all the checkboxes are checked.

Select the Advanced tab in that Properties dialog box. Then scroll down the Property list to locate and select the Wake on Magic Packet property and set its value to Enabled.

Depending on the manufacturer of your PC, you may also need to reboot your computer and access its BIOS settings at boot time. Most PCs allow you to access this at boot by pressing either the Delete or F1 key. The key used to access your computer's BIOS settings at boot time should display briefly on your screen when booting your computer. Look for Power-On or Power Management in your BIOS settings and make sure any options listing Wake on LAN are Enabled.

Enabling Wake on LAN in macOS is much easier. Type in the word *wake* in the search box of the System Preferences dialog. Depending on the version of macOS you're running, this will highlight either the Energy Saver icon or the Battery icon. Select the icon appropriate for your macOS version, then search for and check the Wake for Network Access option. For example, macOS 11 Big Sur users can find this option in the System Preferences -> Battery -> Power Adapter option, as shown in the next screenshot.

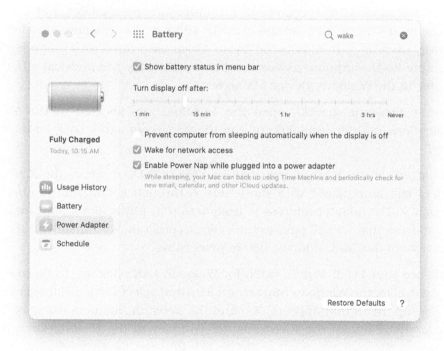

With your target PC or Mac desktop computer properly configured and connected via Ethernet cable to your local area network, along with your Pi on that same network, we can configure the last requirement on your Pi. Open a Terminal window on your Pi and install the wakeonlan program using the apt command:

```
$ sudo apt install wakeonlan
```

Put your Mac or PC desktop to sleep. Windows users can select Sleep from the Windows Start menu -> Power icon, and macOS users can do so by selecting Sleep from the Apple menu.

With the target Mac or PC computer asleep, return to the Terminal window on your Pi and execute the wakeonlan program passing in the MAC address of the Ethernet adapter you captured earlier. Change the letters of the MAC address to lower case, and convert the dashes in between each pair of characters to colons. Remember to replace the placeholder xx pairs in the example below with the actual MAC address values assigned to your Ethernet adapter:

```
$ wakeonlan xx:xx:xx:xx:xx:xx
```

For example, if your Ethernet adapter MAC value is d6:21:b4:3b:a1:c2, then the wakeonlan command would be:

```
$ wakeonlan d6:21:b4:3b:a1:c2
```

Once you've entered the correctly formatted MAC address from your computer in the preferred format and run the wakeonlan program, you should see your Mac or PC turn on. If it failed to do so after twenty seconds or so, check to make sure you're using the correct Ethernet adapter MAC address. If your target desktop computer is Windows, also verify that all that adapter's power settings, including those in the BIOS, have been set correctly. Do not proceed until you can successfully wake up your computer running the wakeonlan command.

With the local area network and computing resources correctly configured, we can continue with the project by establishing free user accounts on the IFTTT and Pushbullet services.

IFTTT

IFTTT, which stands for If This Then That, enables the creation of custom voice-prompted workflows ridiculously easy. This is in comparison to the code and resources required if you opt to set up a stand-alone service of your own. Entire books have been written about doing just that, including *Build Talking Apps for Alexa [Wal22]*. I strongly encourage enthusiastic developers interested in making their own services this way to pursue this path. But for those who want to quickly create a very basic working service, IFTTT offers an easier alternative with just a few clicks of a mouse.

Setting up a free account with IFTTT is simple. Visit ifttt.com on the web and choose the Sign Up button in the upper-right corner of their web home page.

IFTTT allows users to quickly connect Internet of things (IoT) devices such as smart bulbs to web services and smart assistants like Alexa or Google Assistant. Doing so allows you to use a predetermined voice command of your choosing to trigger a cascade of events. For example, you could tell your smart assistant to "feed the dog," which triggers an IFTTT applet to turn on an electric can opener and a pre-programmed internet relay to turn the opened can upside down, dumping the contents into a dog bowl. That sounds like a scene in a movie, doesn't it?

With a new account created and signed in, immediately get to work on creating a new applet by selecting the Create label in the upper-right corner of the page. Select the large If this button and search for Google Assistant (or Amazon Alexa if you prefer that smart assistant).

When you select the Connect button, you'll need to authenticate to the Google (or Amazon) account that's associated with your Google Home (or Amazon Alexa-powered) smart speaker. Please be aware that by allowing IFTTT access to this account, you are granting IFTTT the ability to listen and respond to your voice commands. If you're concerned about this level of privacy, you might want to reconsider even using a smart assistant.

After you Allow IFTTT to link to your Google Home account, complete the IF portion of the IFTTT applet by entering the phrases you want to associate with turning on your computer. IFTTT cannot override existing Google reserved phrases, so you may need to experiment with a set of phrases that works best for your future voice-initiated scenarios. Fortunately, the phrase "Turn on the computer" wasn't a reserved phrase at the time this book was published, so it should work just fine. You can add a few more alternative ways to say this phrase, as well as what you want the Google Assistant to say after it interprets this command. Something like "OK, turning on your computer" should be adequate for this project.

Select the Create Trigger button to save the IF portion of the applet. Now that IFTTT knows what to listen for, you need to tell the service what to do next. Select the Add button in the Then That box and search for the Pushbullet service. Then select the Push a Note from the four different action options, followed by selecting the Connect button as shown in the first screenshot on page 111.

Sign in with the Google account you wish to use to access Pushbullet. This doesn't have to be the same Google account as the one associated with your Google Assistant, but I find it is easier to troubleshoot if it's using the same dedicated account. Just as with Google Assistant, IFTTT needs to be granted access to your Pushbullet account to send it a message, as shown in the second screenshot on page 111.

Enter the message you want Pushbullet to send to the Pushbullet listener script we have yet to create. The title is optional and isn't something we're going to parse, anyway. As a placeholder, you can add Google Voice Command or something more descriptive if you prefer. It's the body of the message that

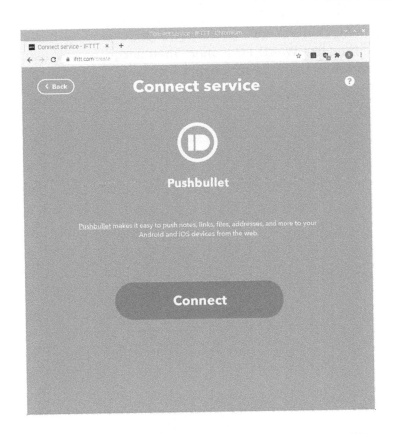

we 'll be parsing in the script, so it's not necessary to make this a long message. Something like WAKE_PC is more than adequate and distinct for our needs, as shown in this next screenshot.

Select Continue to review and finish the applet. When you're satisfied with the defined workflow, select the Finish button to save the IFTTT applet.

Speak to your Google Home or Google Assistant on your smartphone and say "Turn on the computer" to see if IFTTT listens and responds to your request. If not, check your settings before proceeding.

Pushing a Bullet

Sign into pushbullet.com with the same account you used to connect to it during the IFTTT applet creation phase. If you're doing so from the same browser you set up the IFTTT workflow, then Pushbullet will automatically sign you in with that same account.

For a Python script to programmatically listen for Pushbullet messages, it needs an access token that's linked to the account it was generated for. Create this token by navigating to the Settings -> Account page and scrolling down to the Access Tokens section.

Copy this 34-character token and store it in a secure file for now. We'll need that string of characters to authenticate our Python script to listen for any new Pushbullet messages being sent our way.

Scripted WOL

Now that all the back-end messaging infrastructure is configured, we're ready to write the Python script that will receive and process the WAKE_PC message from the body of the Pushbullet message. As with all projects, create an empty Git repository to store and version control your project code:

```
$ git init --bare ~/repository/voicewol.git

$ cd ~/projects/

$ git clone ~/repository/voicewol.git

$ cd voicewol
```

Using your preferred code editor, create a new Python script called wakeup.py. Use the os.system() function to programmatically execute the wakeonlan xx:xx:xx:xx:xx:xx command that you successfully tested earlier:

voicewol/wakeup.py
```
import os
MAC = 'YOUR_COMPUTER_MAC_ADDRESS_GOES_HERE'
os.system('wakeonlan ' + MAC)
```

Put your target desktop to sleep and run the script to verify that it wakes up your computer as expected:

```
$ python3 wakeup.py
```

Your targeted computer should wake up just as it did when you ran the wakeonlan program from the Terminal window. Next up is writing a Python script to listen for and selectively process inbound Pushbullet messages.

Catching a Bullet

Our last major objective for this project is to connect to, listen for, and react to a Pushbullet message containing WAKE_PC in the message body. To do so, we need to install a third-party pushbullet Python library. Unfortunately, the author of this simple pushbullet library failed to properly configure it to be conveniently downloaded and installed using Pip. Instead, it needs to be downloaded directly from the PiPy website. Visit the pypi.org website,[5] search for pushbullet.py, select its link followed by the Download files link on that web page. Download the file ending with the .whl extension, as shown on this screenshot.

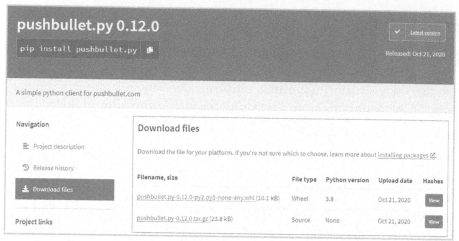

Copy the downloaded pushbullet.py-VERSION_NUMBER-py2.py3-none-any.whl file in your voicewol project directory and install this downloaded file using the pip3 install command. Note that the VERSION_NUMBER in the sample command will be replaced with the latest version the author of the library has deployed:

```
$ sudo pip3 install pushbullet.py-VERSION_NUMBER-py2.py3-none-any.whl
```

Use Pushbullet.py with Python Version 3.6 or Earlier

The version of the Pushbullet.py library used for this project currently only works with Python 3.6 or earlier. You can either install that version of Python on your Pi or try an alternative approach such as Webhooks[6] to connect your smart assistant commands to your Pi. Visit the DevTalk forums to share your experience!

With the pushbullet library installed, we can construct a short Python script that will connect to the Pushbullet service using the access token we registered for earlier. Any time a Pushbullet message is received, we'll print out the body of that message.

Create a new Python script file called pbreceiver.py and add the following code:

voicewol/pbreceiver.py

```
from pushbullet import Listener
from pushbullet import Pushbullet
```

---

5. https://pypi.org/project/pushbullet.py/#files
6. https://ifttt.com/maker_webhooks/details

```
② ACCESS_CODE = 'PUSHBULLET_ACCESS_CODE_GOES_HERE'
  HTTP_PROXY_HOST = None
  HTTP_PROXY_PORT = None

  def on_push(data):
      pbullet = Pushbullet(ACCESS_CODE)
③     pushes = pbullet.get_pushes()

      try:
④         latest = pushes[0]
          print('Latest message body received:\n')
          print(latest['body'])

      except:
          print("No messages.")
⑤ pb = Pushbullet(ACCESS_CODE)

  pbapp = Listener(account=pb,
                   on_push=on_push,
                   http_proxy_host=HTTP_PROXY_HOST,
                   http_proxy_port=HTTP_PROXY_PORT)
  try:
⑥     pbapp.run_forever()
  except KeyboardInterrupt:
      print("Keypress received, exiting script.")
      pbapp.close()
```

Let's take a brief look at what some of the code is doing in this script.

❶ Import the Listener and Pushbullet functions from the pushbullet Python library.

❷ Replace PUSHBULLET_ACCESS_CODE_GOES_HERE with the Pushbullet access code that you previously generated for your login. Set its value to the ACCESS_CODE static variable.

❸ Use this on_push() function to capture pushes received from a Pushbullet event.

❹ Since you're only interested in seeing the first PushBullet message received, only show the body of the Pushbullet message with the print statement. In the final script, we'll be deleting all messages in the queue so as not to reprocess previously received Pushbullet messages.

❺ Here, we initiate the Pushbullet service and assign it to the pb object. Then we pass that object along with the ACCESS_CODE, and any proxy settings if necessary, to the Listener() function. We assign its instantiated result to the pbapp object.

❻ Finally, we instruct the pbapp instance to run_forever(), or at least until we escape out of the program via the Ctrl+Z keyboard combination.

Save and run the pbreceiver.py Python script:

$ python3 pbreceiver.py

With the script running, ask your smart assistant to "Turn on the computer," and a few seconds later, you should see the message body print out on the Terminal:

Latest message body received:

WAKE_PC

If you don't see any message appear, log in to the IFTTT website and double-check the configuration. View the applet's activity to make sure it ran and sent the WAKE_PC text in the body of the Pushbullet note. Verify that the pbreceiver.py script is working correctly before proceeding. You can quit the program by escaping the running script or simply closing the Terminal window it's running in.

Wake Up

We're ready for the moment of truth. Combine the wakeup.py script with the pbreceiver.py script along with a line to clear out all the Pushbullet messages received once the wakeonlan command has been executed. Create a new Python script called voicewol.py and include the following code in it:

```
voicewol/voicewol.py
import os
from pushbullet import Listener
from pushbullet import Pushbullet

ACCESS_CODE = 'PUSHBULLET_ACCESS_CODE_GOES_HERE'
HTTP_PROXY_HOST = None
HTTP_PROXY_PORT = None
MAC = 'YOUR_COMPUTER_MAC_ADDRESS_GOES_HERE'

def on_push(data):
    pbullet = Pushbullet(API_KEY)
    pushes = pbullet.get_pushes()
    try:
        latest = pushes[0]
        if latest['body'] == "WAKE_PC":
            print("Turning on your computer...")
            os.system('wakeonlan ' + MAC)
            pushes = pbullet.delete_pushes()
        else:
            print("Unrecognized code in message body.")
    except:
        print("No messages.")
```

```
pb = Pushbullet(ACCESS_CODE)

pbapp = Listener(account=pb,
                 on_push=on_push,
                 http_proxy_host=HTTP_PROXY_HOST,
                 http_proxy_port=HTTP_PROXY_PORT)
try:
    pbapp.run_forever()
except KeyboardInterrupt:
    print("Keypress received, exiting script.")
    pbapp.close()
```

Save and run the script by passing its name to the Python 3 interpreter.

```
$ python3 voicewol.py
```

Put your target computer to sleep, and then say the magic phrase to wake it back up.

Finally, create a .conf file for Supervisor to consume and use to start the script when the Pi is rebooted and to restart it in case it fails due to an error, low-memory condition, or other unexpected exception:

```
voicewol/voicewol.conf
[program:voicewol]
command = /usr/bin/nohup /usr/bin/python3 \
 /home/pi/projects/voicewol/voicewol.py &
directory = /home/pi/projects/voicewol
user = pi
environment=HOME="/home/pi",USER="pi"
autostart = true
autorestart = true
stdout_logfile = /var/log/supervisor/voicewol.log
stderr_logfile = /var/log/supervisor/voicewol_err.log
```

Save the voicewol.conf file in the /etc/supervisor/conf.d directory, and either restart the Supervisor service or, better yet, reboot your Pi. While you're waiting for the Pi to reboot, put your target desktop computer to sleep:

```
$ sudo reboot
```

Wait a few minutes for your Pi to restart, then say "Turn on the computer" to your smart assistant. Your sleeping desktop computer should wake back up and be ready for your commands.

Awesome! You've earned the convenience of being able to wake up your computer simply by asking your smart voice assistant to do so. After trying it out several times to verify it works consistently, amaze your friends and family with your growing technical know-how and ingenuity by showing how you've harnessed the future of person-to-computer interaction.

Next Steps

By combining the IFTTT and PushBullet services with a smart assistant from Amazon or Google and a short Python script, you can now make your Raspberry Pi turn on your computer with your voice. With this basic IFTTT applet working, you can use your voice to activate other services in your home. Control actuators to unlock and open doors, raise and lower a motorized standing desk, and turn on a sprinkler system—all with the power of your voice. Reverse the message flow to transmit Pushbullet messages to other Internet-connected devices. Install the Pushbullet client on your smartphone and use it in lieu of sending the email alerts we've used in prior projects.

You can also enhance the voicewol.py script to respond to "Turn off the computer" by having your Pi transmit the appropriate command to make your target desktop go to sleep. macOS already has an OpenSSH server installed by default. An OpenSSH server for Windows 10 can be easily installed by selecting Optional Features in the Windows 10 Apps & features Settings page.

The next project will build upon the constructed IFTTT + Pushbullet combo by adding infrared functionality. Doing so will let you activate and control appliances that rely on a legacy line-of-sight, fire-and-forget communication protocol.

CHAPTER 9

Voice IR Control

Smart voice assistants like Amazon Alexa or Google Assistant integrate with a variety of modern home-automation equipment. Network-enabled smart bulbs, outlets, TVs, and entertainment systems that connect to your home network often allow these smart assistants to integrate and control their functionality by voice.

However, there's a world of mostly legacy and low-end consumer electronics that are not network-capable. Instead, they offer the ability to wirelessly interact with the product's controls via InfraRed (IR) signals. This technology has been in the hands of average consumers since the 1970s and is still in use today. IR remote controls are dumb in that they have no confirmation that the signals they transmit actually succeeded in the action they intended to enact. IR transmitters also require direct, unobstructed line-of-sight to the IR receivers that they're transmitting control codes to. But IR still has its uses and continues to be designed into a bevy of consumer electronics. Fortunately, a Pi and a plug-in IR transceiver can bring networked control to these legacy non-networked consumer electronics. And we'll do so without the need for messy breadboards, jumper wires, or other exposed electronics.

Setup

Here's what you need to build this project.

Hardware

- Irdroid transceiver[1]
- Consumer electronic appliance or device with an IR receiver and associated remote control

---

1. https://www.irdroid.com/irdroid-usb-ir-transceiver/

Software

- Linux Infrared Remote Control (LIRC)[2]
- pushbullet[3] Python library

Services

- IFTTT[4]
- Pushbullet[5]

This project builds upon the work we did in Chapter 8, Voice Wake on LAN, on page 105. As such, if you haven't already created accounts on IFTTT and Pushbullet, please revisit the instructions how to do so in the last project. In this project, we'll attach an Irdroid infrared transceiver to the Pi's USB port and then add a few lines of Python code to the previously completed voicewol.py script. Doing so will smarten up the ability to interact with a chosen IR-controlled device, using just your voice.

Before attaching the Irdroid IR transceiver to the Pi, install the Linux Infrared Remote Control (LIRC) service and helper apps for the Pi:

```
$ sudo apt install lirc
```

Next, plug the Irdroid transceiver into one of the powered Pi's available USB ports, as shown in the next photo. You should see Irdroid's internal blue LED flicker once or twice to acknowledge its USB connection.

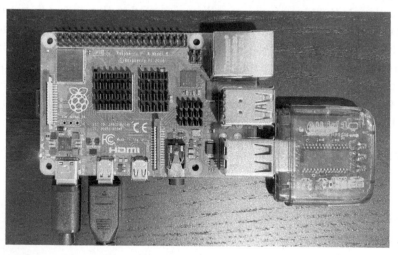

---

2. https://lirc.org/
3. https://github.com/rbrcsk/pushbullet.py
4. https://ifttt.com/
5. https://www.pushbullet.com/

Connecting Irdroid to Linux Kernels Higher Than 5.7

 The Irdroid instructions in this chapter are for Linux kernels 5.7 or less. If you are running a version of Raspberry OS with a higher kernel version, simply plug the Irdroid into the Pi's USB port, set the device to /etc/lirc0, and leave the driver assignment to default in the lircd.conf file.[6] After a reboot, your Irdroid should be responding and ready to go!

Verify the Irdroid hardware is indeed recognized by the Raspberry Pi OS. Look for a device named ttyACM0 in the Pi OS's root level /dev directory. Note that if you happen to have another USB device connected that also uses ACM, the Pi may reassign the Irdroid hardware as ttyACM1 or additional increments. If you want to be absolutely sure the Irdroid is recognized, remove all other USB devices before connecting the Irdroid. You can confirm the Irdroid device is connected by issuing the following statement in the Pi's Terminal window:

```
$ ls /dev/ttyACM*
```

If successful, the Terminal will echo the path, /dev/ttyACM0. If not found, the ls command results will complain that no such file or directory exists. It's very important that the device is recognized before proceeding.

Now edit the LIRC options configuration file to tell LIRC what type of driver to use to talk to the Irdroid as well as Irdroid's device location assignment:

```
$ sudo nano /etc/lirc/lirc_options.conf
```

Locate the driver and device entries in the file and replace their assignments with irtoy and /dev/ttyACM0, respectively:

```
[lircd] nodaemon = False
driver = irtoy
device = /dev/ttyACM0
```

Save the assignment changes made to the lirc_options.conf file and reboot the Pi:

```
$ sudo reboot
```

Once the Pi is back online, open a Terminal window and test that the Irdroid's receiver is recognizing inbound IR transmissions. Using a working infrared remote control from the consumer electronic device you intend to remotely control with the Irdroid-assisted Pi, launch the mode2 app with the following command:

```
$ mode2 --device /dev/ttyACM0
```

---

6.   https://irdroid.eu/irdroid-usb-infrared-transceiver-linux-kernels-greater-5-8/

If the Irdroid has been registered correctly with the irtoy driver, the mode2 app should respond with the following output:

```
Using driver irtoy on device /dev/ttyACM0
Trying device: /dev/ttyACM0
Using device: /dev/ttyACM0
```

Now point your chosen, working IR remote at the Irdroid and press any of the remote control's buttons. You should see the Terminal window output a series of numbers alternating between pulse XXXX and space XXXX where XXXX represents a series of integers. If you don't see any results on your screen, make sure the IR remote you're using is working with its associated consumer electronic. Also be sure to point the remote's IR transmitter directly at the Irdroid, since IR requires line-of-sight to work. When you're done testing the remote, press the Ctrl+C keys to escape the mode2 program or close the Terminal window running it.

We're nearly done with configuring LIRC. The only major configuration you need to make is to tell it what kind of remote control you're using so LIRC knows the correct remote control IR codes to transmit when asking it to perform a function. If your IR device you want to control is a popular TV, DVD player, or stereo receiver, someone may have already done the work of recording the IR control codes for you. You can browse the LIRC Remotes Database[7] to see if your device is listed. If it is, download its related .conf file and save it in the /etc/lirc/lirc.conf.d/ directory. You can do this for all the IR remote-controlled devices you wish to programmatically operate from the Pi. This is because LIRC's remote conf files each have their own unique name to identify the remote codes to be transmitted.

If your particular IR remote-controlled TV or other device isn't listed in the LIRC Remotes Database, you can use Irdroid's built-in IR receiver to record the IR codes emitted from your desired remote control. For example, I have an old Logitech DTS decoder that powers the audio emitted from my PC's sound card. Since no such configuration file for this device existed in the LIRC Remotes Database, I used irrecord to capture and store its remote codes. With your remote control in hand, run LIRC's irrecord program in a Terminal window:

```
$ irrecord --disable-namespace
```

Follow the program's onscreen instructions, including giving the remote-controlled device a name. For example, I called my remote recorded device dts and this saved my captured IR codes to a dts.lircd.conf file in the Pi's home directory. The automatically generated configuration file will have a different name depending on what you recorded. The irrecord program will ask you to

---

press all the remote's buttons in various random ways. When it has captured enough of a baseline, it'll be able to assign specific names to the related remote-control buttons being pressed. Name each one accordingly. For example, I named the power button on my DTS remote control power, and so forth. Once you've used irrecord to capture all the remote's buttons, follow the program's instructions to exit and save the configuration file. Assuming you executed the irrecord program from the Pi's default home directory, you should see the newly generated NAME_OF_REMOTE.lircd.conf file in that directory where NAME_OF_REMOTE is the name you assigned during the recording sequence. Copy this generated configuration file into LIRC's configuration directory:

```
$ sudo cp ./NAME_OF_REMOTE.lircd.conf /etc/lirc/lircd.conf.d
```

For LIRC to recognize this new configuration file, you can either restart the LIRC daemon or reboot the Pi. I prefer a reboot since it's quick and also verifies that the configuration changes made are permanent.

Test It

The moment of truth has arrived. It's time to see if all that time invested in configuring LIRC has paid off. Make sure the IR receiver of the consumer electronic device you captured the IR codes for is in direct line of sight of the Irdroid. Then open a Terminal window and, assuming you recorded a power-button command during your irrecord session, issue an irsend command to turn on the device. In the case of my Logitech DTS receiver, I issued the following command to turn it on or off:

```
$ irsend SEND_ONCE dts power
```

If everything works, you should see your IR-controlled device come to life. Issue other commands you named and recorded with irrecord using the irsend program. When you are satisfied with how easy it is to remotely control your IR device from the Pi, incorporate the irsend command into Python scripts using the os.system() call to trigger these codes programmatically. For example, here's a line of Python that turns on my old Philips HDTV, courtesy of the Philips_RC-7843 configuration file contributed to the LIRC Remotes Database by Armenio Castro Vale:

```
os.system("irsend SEND_ONCE PHILIPS_RC-7843 KEY_POWER")
```

You can even write a simple Python script to execute multiple IR send commands. Just be sure to give enough time in between each transmission to wait for the code to transmit and enable the state change on the receiving device.

Add It

Cool, we can now programmatically control legacy infrared-receiving consumer electronics. It's time to incorporate that single line of irsend code into the voicewol.py Python script we created and configured in the Voice Wake on LAN project. You can either associate it with the existing WAKE_PC body tag we associated with waking the PC or create a new IFTTT/Pushbullet sequence to control a device separate from the Wake PC event trigger. For example, I created a separate body message called PWR_TV to simulate pressing the power button on my Philips TV. Then I created a new IFTTT applet, following the same steps described in the Voice Wake on LAN project, calling this new applet "Turn on/off TV", and assigned a new phrase. I couldn't use the phrase "Turn on the TV" because it was too generic for Google to parse. Instead, I more specifically used the phrase "Turn on the Philips," which was more specific for Google to translate and properly route through IFTTT. For the Pushbullet message, I added the aforementioned PWR_TV label. Here's the final modified voicewol-ir.py script incorporating these additions:

```
voiceir/voicewol-ir.py
import os
from pushbullet import Listener
from pushbullet import Pushbullet

ACCESS_CODE = 'PUSHBULLET_ACCESS_CODE_GOES_HERE'
HTTP_PROXY_HOST = None
HTTP_PROXY_PORT = None
MAC = 'YOUR_COMPUTER_MAC_ADDRESS_GOES_HERE'

def on_push(data):
    pbullet = Pushbullet(API_KEY)
    pushes = pbullet.get_pushes()
    try:
        latest = pushes[0]
        if latest['body'] == "WAKE_PC":
            print("Turning on your computer...")
            os.system('wakeonlan ' + MAC)
            pushes = pbullet.delete_pushes()
            os.system('irsend SEND_ONCE dts power')
        elif latest['body'] == "PWR_TV":
            print("Pressing TV remote power button")
            os.system('irsend SEND_ONCE PHILIPS_RC-7843 KEY_POWER')
        else:
            print("Unrecognized code in message body.")
    except:
        print("No messages.")

pb = Pushbullet(ACCESS_CODE)

pbapp = Listener(account=pb,
```

```
                on_push=on_push,
                http_proxy_host=HTTP_PROXY_HOST,
                http_proxy_port=HTTP_PROXY_PORT)
try:
    pbapp.run_forever()
except KeyboardInterrupt:
    print("Keypress received, exiting script.")
    pbapp.close()
```

Create your own sequence of IR-controllable actions using a variety of sequences and voice commands. Do keep in mind that IFTTT limits users subscribed to its free account tier a maximum of three applets. If you need more, you can upgrade to their paid Pro subscription plan.

Blast It

Update the path and name of the modified Python script in the related voice-wol.conf supervisor file from the prior Voice Wake on LAN project. Restart the Supervisor service or reboot the Pi. Ask Google or Alexa to "Turn on the PC" or whatever other command you associated with your IR-controlled device, and your device should respond accordingly. If you encounter any issues, troubleshoot by making sure you can communicate with your device directly from the irsend command in the Terminal.

Nice job! You've just brought your previously dumb, non-networked consumer electronic device of the past into the smart, networked world of the present. You can also put your old remote in a junk drawer, making sure to remove the batteries so they don't leak in the remote after years of forgotten disuse.

Next Steps

You've seen how easy it is to build upon the foundation of past projects. Besides the time it took to configure the Irdroid with the Pi, adding your IR-controlled device to an existing voice project literally took minutes to enable. You also have the benefit of adding more legacy IR control systems to your Pi and triggering those by voice or other means. For example, if you have a ceiling fan controlled by an IR remote, you can incorporate its associated irsend command with the temperature-triggering code from Chapter 5, Hue Fan, on page 61, to turn the ceiling fan on and off, depending on the warmth in the room. Some window blinds are controlled via infrared triggers, and these could be combined with the code from Chapter 6, Hue Auto Light, on page 75, to raise and lower, based on the amount of outdoor light or time of day.

You can also add other infrared devices in the same line of sight as the device you're managing. Imagine a symphony of actions instantiated by a single voice command. If you choose to do so, be aware that it's best that the other devices are designed and manufactured by entirely different companies. It's unlikely that those other devices will use the same IR codes to activate their device's customized features.

The next project will return to giving the Pi something else to do besides interacting with hardware. We're going to build an autonomous software bot to save time retrieving and acting on timely information that's relevant to you.

CHAPTER 10

RedditBot

Throughout this book, we've explored projects that leverage the Pi hardware in terms of both its ability to interact with sensors and actuators and its capability to function as a low-power, always-on server. This project further accentuates the latter by turning a popular, albeit relatively static, web experience into an interactive chatbot.

I occasionally call upon Reddit to explore my interests, spanning across programming and DevOps to board and video games. One of the persistent Reddit presentation themes that never really stuck with me is that, unlike email and RSS feeds, it's not easy from Reddit's web interface to bulk hide read messages on Reddit. I'm an Inbox Zero[1] practitioner, and seeing persistent read messages onscreen torques my organizational sensibilities. Fortunately, Reddit provides developers access to its API, making it easy to create your own data interface. By combining the Reddit API with the popular and very easy-to-program Discord chatbot interface, I can have the best of both worlds. And if you follow along and complete this project, so can you.

Setup

Here's what you need to build this project.

Hardware

- Smartphone, Mac, or PC running the official Discord app

---

1. https://www.43folders.com/izero

Software

- Async PRAW:[2] the Asynchronous Python Reddit API Wrapper
- Discord[3] Python library

Services

- Discord[4] user account (free)
- Reddit[5] user account (free)

Install the two necessary Python library dependencies for the project.

```
$ sudo pip3 install asyncpraw
```

```
$ sudo pip3 install discord
```

With the Python dependencies installed, we're ready to start coding.

Reddit Access

Log in to your Reddit account via a web browser and, if you haven't done so already, subscribe to a couple subreddit topics you're interested in. Use the Search field on the Reddit home page to find subreddit topics you may be interested in following. For example, if you're interested in following fellow Redditor posts on Python, join the /r/Python[6] subreddit.

After you've subscribed to a couple choice subreddits, visit the apps tab[7] on your account's Preferences web page and click the create app... button at the bottom of that page. Give your app a name, mark it as a script, and set the redirect URI to http://localhost:8080.

Submit your app definition and copy the 14-character autogenerated client ID, located under the saved app name, and the 30-character generated secret. You'll need these two strings to authenticate to Reddit and authorize the script's access to your account data.

With your client ID, secret, Reddit username, and account password in hand, create a new Python script called reddittest.py, using your preferred code editor, and copy the following code into it. Be sure to replace the four GOES_HERE placeholders with the respective values:

---

2.   https://pypi.org/project/asyncpraw/
3.   https://pypi.org/project/discord.py/
4.   https://discord.com/
5.   https://www.reddit.com/
6.   https://www.reddit.com/r/Python/
7.   https://www.reddit.com/prefs/apps/

```
redditbot/reddittest.py
import asyncio
import asyncpraw

async def main():

    reddit = asyncpraw.Reddit(client_id="APP_CLIENT_ID_GOES_HERE",
            client_secret="APP_CLIENT_SECRET_GOES_HERE",
            password="REDDIT_USER_PASSWORD_GOES_HERE",
            user_agent="RedditBot",
            username="REDDIT_USERNAME_GOES_HERE")

    print('Your Reddit username is...')
    print(await reddit.user.me())

    await reddit.close()

if __name__ == "__main__":
    loop = asyncio.get_event_loop()
    loop.run_until_complete(main())
```

Save and execute the script:

$ python3 reddittest.py

Verify that it successfully prints out your Reddit username. If it fails, make sure your app and user account credentials were correctly copied without spaces in the GOES_HERE placeholder text. Also confirm that your Pi is connected to the Internet.

Notice when the script runs successfully the Terminal will initially print out Your Reddit username is... and wait a few seconds before finally printing out your username. That's because we're making a nonblocking asynchronous call to authenticate to and retrieve data from Reddit. That's also why the structure of this script looks different from any others we've run thus far in the book.

We could have used the standard PRAW[8] library instead of the new and improved Async PRAW, but because we need the functionality of PRAW in a Discord bot, the results awaiting to be received and displayed by our bot must be asynchronous. This allows our bot to continue to do things while it waits for results to return from our calls to Reddit.

Depending on server load, number of submissions, amount of data returned, and other factors, this can be right away or take several or more seconds to complete. If we didn't make these function calls asynchronous, our script would come to a standstill, waiting for all the results to be received before executing any other instructions. Doing so would make our bot unresponsive and, in a worst case scenario, crash entirely.

___

8. https://pypi.org/project/praw/

Using asynchronous libraries in Python requires additional planning and a deeper understanding of Python's async and await keywords. Fortunately the scripts we need to complete this project are easy to read and use without having to delve into the depths of Python's asynchronous calls. But if you intend on building more complex bots or scripts that need to keep running while waiting for results from a third-party web service, invest the time to learn more[9] about asynchronous programming in Python 3.7 and higher.

Replace Passwords with Refresh Tokens

To maintain a good security standard and give developers the ability to revoke compromised security keys without having to delete their user accounts, Reddit employs the use of authentication tokens in lieu of usernames and passwords to authenticate applications and their users to a service.

If you intend to share your script's functionality with others, or are concerned about passing your cleartext Reddit username and password through the various Python libraries that Async PRAW uses, consider creating a *refresh token*. This unique secret takes a bit more work up front to generate, but doing so makes your Reddit applications more secure.

Should the token ever get compromised for any reason, simply regenerate a new one; there's no need to delete your user account, although it still would be good security practice to change your account password any time a security breach is suspected. While the generation of a refresh token is beyond the scope of this book, if you're security conscious, I strongly recommend that you pursue the topic further by reading the Async PRAW documentation.[a]

---

a.   https://asyncpraw.readthedocs.io/en/latest/getting_started/authentication.html

Now that we've successfully authenticated to Reddit within a Python script, we can build upon that authentication to retrieve and interact with a variety of submissions posted by fellow Redditors in our account's subscribed subreddits.

Reddit Commands

The Reddit API[10] provides a wealth of different function calls that can perform a number of actions, ranging from subscribing and unsubscribing to subreddits to posting new submissions, emojis, and flairs. Specifically for this project, we want to retrieve only those subscribed subreddit submissions that are new posts. Once retrieved and read, we'll also want to hide those posts so they don't keep showing up in our new Reddit submission queries.

---

9.   https://docs.python.org/3/library/asyncio.html

10.   https://www.reddit.com/dev/api

Show Titles of New Reddit Posts

Async PRAW makes it easy to interact with Reddit content. Thanks to this powerful Python library, we can iterate through all the recent submissions posted to a subreddit in just a few lines of code. Do this now by creating a new Python script file called reddlist.py and populating it with the following code:

```
redditbot/reddlist.py
import asyncio
import asyncpraw

async def main():
    reddit = asyncpraw.Reddit(client_id="APP_CLIENT_ID_GOES_HERE",
            client_secret="APP_CLIENT_SECRET_GOES_HERE",
            password="REDDIT_USER_PASSWORD_GOES_HERE",
            user_agent="RedditBot",
            username="REDDIT_USERNAME_GOES_HERE")
    async for mysubreddit in reddit.user.subreddits(limit=None):
        current = mysubreddit.display_name
        print('Subreddit: ' + current)
        subreddit = await reddit.subreddit(current)
        async for submission in subreddit.new(limit=20):
                print(submission.title)

    await reddit.close()

if __name__ == "__main__":
    loop = asyncio.get_event_loop()
    loop.run_until_complete(main())
```

Let's take a closer look at the instructions issued to retrieve the list of subscribed subreddits and then iterate through each of them for a list of new posts.

❶ Use a for loop to iterate through each of the current user's subscribed subreddits.

❷ Print the current subreddit's name being processed each time a subscribed subbreddit is iterated upon.

❸ Likewise, iterate through each of the new submissions posted to the current subreddit being iterated upon. Notice that we're limiting the total number of new posts being processed to 20. This amount can be maximized up to 50 new posts, but it will take longer to process the retrieval of the article titles. The more subreddits your Reddit user account is subscribed to, the longer it will take for this script to execute overall.

Save and execute the scriipt:

```
$ python3 redditlist.py
```

You can verify that the titles of the new submissions being displayed in the Terminal window are identical to those on your logged-in Reddit home page's new tab.

This is nifty, but unless the subreddit is wildly active, many of the same items will continue to show up on the list if you run the script several times a few minutes apart. This makes parsing what is new since the last time the script was run a cognitive burden. We could employ a database approach similar to what was done in previous chapters, but an easier and cleaner approach would be to simply hide those viewed posts so they don't continue to reappear each time a new submission inquiry is requested.

Hide New Reddit Posts

Create a new Python script called reddhide.py and copy the following code into it:

```
redditbot/reddhide.py
import asyncio
import asyncpraw

async def main():

    reddit = asyncpraw.Reddit(client_id="APP_CLIENT_ID_GOES_HERE",
            client_secret="APP_CLIENT_SECRET_GOES_HERE",
            password="REDDIT_USER_PASSWORD_GOES_HERE",
            user_agent="RedditBot",
            username="REDDIT_USERNAME_GOES_HERE")

    async for mysubreddit in reddit.user.subreddits(limit=None):
        current = mysubreddit.display_name
        subreddit = await reddit.subreddit(current)
        async for submission in subreddit.new(limit=20):
            if (submission.hidden is False):
                print('Hiding ' + submission.title)
                await submission.hide()

    await reddit.close()

if __name__ == "__main__":
    loop = asyncio.get_event_loop()
    loop.run_until_complete(main())
```

The intialization and iteration code are identical to the redditlist.py script, but note the addition of the segment of code that checks for a hidden condition and then hides the submission if it isn't hidden already.

```
async for submission in subreddit.new(limit=20):
    if (submission.hidden is False):
        print('Hiding ' + submission.title)
        await submission.hide()
```

Once again, we're limiting the number of new submissions being iterated upon to 20. If the submission being queried is visible (in other words, it's hidden value is False), then inform the user by printing to the Terminal that it is being hidden and then wait for that submission to actually be set to hidden.

Save and run the script in the usual way:

```
$ python3 reddhide.py
```

As each new visible submission is processed, it'll be displayed accordingly and then set to hidden. Refresh the new tab on the user account's Reddit home page and you should see those new submissions processed by the reddhide.py script are no longer showing up on that web page. If you run the reddhide.py script again, only those new submissions posted since the last time you ran the script will be displayed. They'll also be marked as hidden.

Now that we know how to programmatically list and hide new submissions in subscribed subreddits, we can transplant this code into Discord bot code that can transform this bland Terminal output experience into a fun, interactive one.

Bot Creation

As the Internet generation adopts virtual meetings as a mainstay of modern times, services like Zoom, Teams, and Discord have exploded in popularity. Yet unlike Zoom and Teams, which are focused on education and the enterprise, Discord doesn't take itself too seriously. Consequently, it's also by far the easiest of these platforms to create an interactive bot using the discord.py Discord Python library and a few lines of code.

If you haven't already done so, create and log in to[11] your Discord account on the web. Discord provides each authenticated user their own server. The server's name is the same as the logged-in username. Play around with the interface, take a look at the list of Text Channels (Discord automatically creates a #general channel for all new users) and post a few messages to yourself in any active channel on your Discord server. After you've spent some time acclimating to the Discord interface, proceed with the creation of a new Discord bot.

While you're still logged in to your Discord account within a web browser, visit the developer's application page[12] and select the New Application button.

Name your new application RedditBot and select the Create button.

---

11. https://discord.com/login
12. https://discord.com/developers/applications

Feel free to assign additional details, such as an icon for your new application to use and a brief description summarizing what the bot does. Then on the left margin of the page, select the Bot option. This will display another overlay asking you to confirm your intent to Build-A-Bot.

Acknowledge the Add A Bot To This App pop-up alert to add this bot to the RedditBot app. Be aware as the alert indicates, this is an irrevocable action. Select the button labeled "Yes, do it!" to proceed.

Now that your application's bot has been brought to life, you should see to the right of your bot's icon (which by default is the same image that you assigned earlier to your application) a Token label. Select the Copy button to copy the token. You'll need this to programatically bring your bot to life in your application. If for any reason you believe the token was compromised, you can select the Regenerate button to expire the old token and assign a new one to the bot. After the token has been copied, store it for use in the upcoming Python script.

Next, select the OAuth2 tab on your RedditBot's application page and check the bot settings to indicate that this app is indeed a Discord bot.

Scroll down to the Bot Permissions needed for RedditBot to interact with you on your Discord server. Limit the scope of permissions to only those that you know the bot needs. RedditBot needs to send messages with embedded web links. And since we want to practice Inbox Zero even with RedditBot interactions within its dedicated Discord channel, it will also need access to Message History so we can ask RedditBot to delete old messages when we prefer to do so. Under Text Permissions, check the boxes for Send Messages, Send TTS Messages, Manage Messages, Embed Links, and Read Message History.

Scroll back up to the Scopes box and Copy the URL generated in this section. Open a tab on the same browser you logged in to Discord with and paste the generated URL into that new tab. Acknowledge and authorize RedditBot to access your Discord channels.

Confirm and approve to RedditBot the ability to Send Messages, Send TTS Messages, Manage Messages, Embed Links, and Read Message History in the messages it posts in your Discord channel(s). These permissions should be identical to those you defined in the previous bot's permissions screen.

Once your custom RedditBot has been authorized, Discord will respond with an approval message.

Awesome! You now have an inactive RedditBot assigned to your Discord server. What I prefer to do with my new bots, especially those that are only supposed to be used by me and no other subscribers to my server, is create a new, dedicated text channel on my Discord server. I then assign my bot to that dedicated channel and remove the ability for the bot to interact on any other channel except that specifically dedicated one.

To limit the new RedditBot to a dedicated channel as opposed to all your account's Discord channels, go to your Discord desktop app or web interface and create a new text channel by selecting the + icon next to the Text Channels column on the left side of the app. Name the new text channel #reddit. Make it private for now, and exclusively grant your RedditBot access to it.

Next, limit the RedditBot's permissions on all other channels in your Discord account, such as the #general channel the new bot was automatically assigned to by default. Edit the channel by selecting the sprocket icon next to the channel's name, then select the Permissions tab and highlight the RedditBot. Deactivate all channel permissions for that channel, so that your RedditBot is no longer visible in that channel's Members column on the right side of the screen.

After you've set the permissions for all other channels along with granting RedditBot access to your new #reddit channel, you should only see the currently offline RedditBot visible in that channel.

Visit any other channels on your Discord server, and confirm that you cannot see your offline RedditBot in any of your other Discord account channels.

Now that you've configured your dedicated #reddit channel that only you and your RedditBot have access to, it's time to bring the RedditBot online with a little help from the Discord Python library.

Bot Interaction

Building Discord bots using the discord Python library is a simple affair. After importing the discord library into your Python script, you just need to instantiate a bot object and pass it the token that was generated during your bot's application configuration. The discord Python library utilizes a cool Python feature called decorators[13] that acts like a function or class, wrapping another function or class and changing its behavior. Not only does this allow chaining of multiple functions and classes, but the decorator syntax keeps Python code from bloating with unnecessary complexity and verbosity.

---

13. https://wiki.python.org/moin/PythonDecorators

For our Discord RedditBot, we need it to respond to three different inputs:

- !list
- !hide
- !clear

!list will call the subreddit submission title routine that we wrote earlier. Additionally, to make this result more useful, we'll embed a web link to the web page referenced in the submission. This will allow the RedditBot to highlight title results with a corresponding hyperlink.

!hide will call the subreddit hide submission routine that we previously wrote and tested. When this command is issued, the most recent twenty new submissions will be marked as hidden. That way, they will no longer appear in subsequent !list requests.

!clear extends the Inbox Zero practice to the Discord channel. Those who prefer to leave old Discord posts persistent in their channels are likely the same people who don't mind seeing tens of thousands of read emails in their inboxes. While you may choose to never call upon this function, it will be there in case you need to clean up your #reddit channel.

Using decorators, the Python discord library allows the simple creation of these commands and assigns them to the instantiated bot object. Additional parameters for command aliases and help messages can be included. For instance, it would be tiresome after using RedditBot for awhile to keep having to type out !clear every time the channel needed to be cleared of past messages. Assigning an alias like !c would be much faster and accomplish the same objective.

With these basic guidelines in mind, you can create a new bot command using the @bot.command decorator. Here's the bot command assignment for the list command:

```
@bot.command(name='list', aliases=['l'], \
help='List new posts in subscribed reddits.')
```

Once a command has been defined, it needs to have a corresponding function to call upon while passing in the context of the channel the bot is interacting with. For example, here's the reddlist() function copied from the reddlist.py script we wrote earlier, assigned to the !list @bot.command():

```
async def reddlist(ctx):
    async for mysubreddit in reddit.user.subreddits(limit=None):
        current = mysubreddit.display_name
        await ctx.send('Checking ' + current + '...')
        subreddit = await reddit.subreddit(current)
```

```
        async for submission in subreddit.new(limit=20):
                embed = Embed(title=submission.title, url=submission.url)
                await ctx.send(embed=embed)
```

Let's apply this approach to the remaining bot commands. Create a new red-ditbot.py script in your preferred code editor and include the following lines of Python code:

redditbot/redditbot.py
```python
import asyncpraw
from discord import Embed
from discord.ext import commands

BOT_TOKEN = "BOT_TOKEN_GOES_HERE"

reddit = asyncpraw.Reddit(client_id="APP_CLIENT_ID_GOES_HERE",
            client_secret="APP_CLIENT_SECRET_GOES_HERE",
            password="REDDIT_USER_PASSWORD_GOES_HERE",
            user_agent="RedditBot",
            username="REDDIT_USERNAME_GOES_HERE")
```

❶ `bot = commands.Bot('!')`

❷
```python
@bot.command(name='list', aliases=['l'], \
help='List new posts in subscribed reddits.')
async def reddlist(ctx):
    async for mysubreddit in reddit.user.subreddits(limit=None):
        current = mysubreddit.display_name
        await ctx.send('Checking ' + current + '...')
        subreddit = await reddit.subreddit(current)
        async for submission in subreddit.new(limit=20):
                embed = Embed(title=submission.title, url=submission.url)
                await ctx.send(embed=embed)

@bot.command(name='hide', aliases=['h'], \
help='Hide new posts in subscribed reddits.')
@commands.has_role('Announcements')
async def reddlist(ctx):
    async for mysubreddit in reddit.user.subreddits(limit=None):
        subreddit = await reddit.subreddit(mysubreddit.display_name)
        async for submission in subreddit.new(limit=20):
                await submission.hide()
                await ctx.send('Hidden: ' + submission.title)
```

❸
```python
@bot.command(name='clear', aliases=['c', 'delete', 'd'],
            help='Clear up to 100 messages at a time in the \
            current channel.')
async def clear(ctx):
    await ctx.channel.purge(limit=100)
    await ctx.send('Messages deleted.', delete_after=4)
```

❹
```python
@bot.event
async def on_command_error(ctx, error):
```

```
    if isinstance(error, commands.errors.CommandNotFound):
        await ctx.send("I don't understand that command.")
```

```
bot.run(BOT_TOKEN)
```

Let's do a quick review of some of the new code fragments in this final version of the RedditBot codebase.

❶ Initialize the bot object and assign the exclamation point character as the identifier to precede a bot command. You can change this character to anything you prefer, but the common Discord bot practice is to use an exclamation point.

❷ The routines for listing new subreddit submissions and hiding those submissions were copied straight from the previously written reddlist.py and reddhide.py scripts.

❸ For the !clear bot command, the channel.purge function is called, with a limit of 100 prior text messages to be searched through and potentially removed. Only messages that are less than two weeks old will be deleted when this command is issued.

❹ The @bot.event responds to events that may occur, such as errors. In the event that the bot receives an undefined command such as using !show in place of !list, the bot responds with I don't understand that command.

Save the project code and execute the script:

```
$ python3 redditbot.py
```

Go to your Discord desktop, smartphone, or web app and visit the #reddit channel. You should now see your RedditBot online and ready for interaction. Refresh a list of commands you can submit to it by sending a !help message. Pulling from the help statements we included in the respective command's initialization, RedditBot should respond with a list of commands it knows how to process. Send it a !list command. If any new, unhidden submissions are available, RedditBot will list those submission titles along with hyperlinks to the web page referenced by it. Issue a !hide command to hide those submissions from showing up again in a future !list request. Lastly, submit a !clear command to clear out the text content that was posted in your #reddit text channel. That single command is so much easier than manually right-clicking every individual line and selecting Delete message from the corresponding pop-up menu.

I really like the way the Messages deleted entry displayed by RedditBot cleans up after itself by deleting its own message four seconds after it posts, as defined in this line of code:

```
ctx.send('Messages deleted.', delete_after=4)
```

Fantastic job on creating your own custom RedditBot! All that remains to do is set up the script in Supervisor so that it will automatically run when your Pi starts up and stay running in the background, ready to process your next Reddit query at any time:

redditbot/redditbot.conf
```
[program:redditbot]
directory = /home/YOURUSERNAME/projects/redditbot
command = /usr/bin/nohup /usr/bin/python3 redditbot.py &
user = YOURUSERNAME
environment=HOME="/home/YOURUSERNAME",USER="YOURUSERNAME"
autostart = true
autorestart = true
stdout_logfile = /var/log/supervisor/redditbot.log
stderr_logfile = /var/log/supervisor/redditbot.log
```

Restart the Supervisor service or reboot your Pi to run the script. Verify that your RedditBot is online using any Discord app (desktop, mobile, web) you prefer, and command away!

Next Steps

Great job on creating your very own Discord bot that will respond to queries about subreddits that you're interested in following. You can use the Discord client for checking in on your subreddits at any time!

Now that you have the basics down on creating Discord bots in Python, explore the asyncpraw and discord libraries more thoroughly. Doing so will give you a deeper appreciation for all the hard work that the authors of these libraries have done as well as fuel your creativity by extending your RedditBot to do more. Here are just a few ideas to expand upon to improve your RedditBot.

Convert the RedditBot to monitor an asyncpraw subreddit stream to automatically text you new Reddit submissions as soon as they are posted to Reddit. Add the ability to view, reply, and post new subreddit comments directly within the Discord client. Inquire about other Redditors, display your own Reddit stats, vote up or down submissions, and much more.

In the next and final project, we will leverage the Discord Python library to allow us to post real-time motion-detected image snapshots to a dedicated Discord channel. In a way, it will be a compact culmination of a number of approaches learned while building other projects in this book.

CHAPTER 11

PhotoHook

In this project, we'll create a Python script that will transmit a notification with an accompanying photo whenever motion is detected near the location of the Pi.

Prior projects in this book have relied on email messages as the predominant way to receive event notifications. But email is *so* twentieth century. It offers a tried-and-true, pull-to-retrieve message mechanism that works best for long-form messages benefiting from permanent record storage. But for the types of messages that just need a quick alert, email is slow and heavy. Something like a desktop or smartphone notification push message would be more beneficial. Fortunately, we already used a service that has a robust client capable of displaying notifications on computers, phones, and tablets. In addition to Discord offering developers an easy way to build interactive bots, the service also allows for the receipt of notifications via webhooks.[1]

While we could expand upon RedditBot to bundle Photo alert functionality into it, doing so would require more sophisticated use of Python's thread model. Recall that Discord bots operate their own asynchronous event loop. Without setting up a separate managed thread to monitor sensor activity, we could come into conflict with the While loops we used in those monitoring scripts.

Fortunately, Discord offers a very easy solution to this problem: webhooks. Webhooks allow us to post any message to a channel with minimal overhead and without requiring a dedicated bot. In fact, webhooks are so lightweight that they don't even require any third-party Python libraries. Just a call to Python's built-in requests library will do fine. Also, we can swap out the email code we used in the sensor projects and replace it with a few lines of Discord

---

1. https://discord.com/developers/docs/resources/webhook

webhook code. Since we already built the Hue AutoLight project that used the SensorHub's motion detector, we already have a majority of the script written. Let's do this!

Setup

Here's what you need to build this project.

Hardware

- Docker SensorHub
- Pi Camera v2 or USB webcam

Software

- Raspistill[2] Pi Cam software included in Raspberry Pi OS

Services

- Discord[3] user account (free)

Begin the project like most others, with the installation of any necessary third-party Python libraries. If you didn't build the RedditBot project, be sure to install the Discord Python library:

```
$ sudo pip3 install discord
```

Also, if you currently have the SensorHub still attached to your Pi from the Hue Fan or Hue AutoLight projects, shut down your Pi and temporarily remove the SensorHub to install the Pi Camera. You'll reattach the SensorHub once the Pi Camera is connected to the Raspberry Pi's camera port.

Camera Attachment

Look for the camera port on the top of the Pi. On the Pi 4 Model B, this port is located to the left of the USB ports. Gently pull up on the plastic clip that will be snapped back into place once the Pi Camera ribbon is seated inside the clip.

With the blue edge strip of the Pi Camera ribbon facing the USB ports, slide the ribbon into the Pi camera port until it is flush against the base of the Pi. Be sure to seat the ribbon in straight and don't crimp it or squash it into the clip slot. Refer to the first photo on page 143 showing the correct orientation and seating of the ribbon.

2.　https://www.raspberrypi.org/documentation/raspbian/applications/camera.md
3.　https://discord.com/

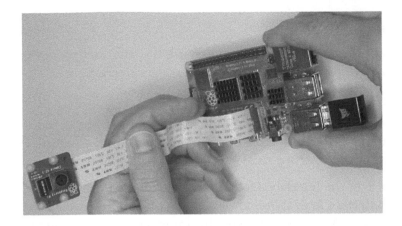

When the camera ribbon is properly seated, lock it in place by pressing down on the Pi's ribbon clip. You may want to use both thumb and forefinger to apply equal pressure across the clip when pressing down. Again, pay special care to seat the ribbon in the clip flat so it's not angled, and thus confirm all ribbon metal contacts align with the contacts on the Pi's camera port. Refer to the next photo for proper alignment of the camera ribbon cable.

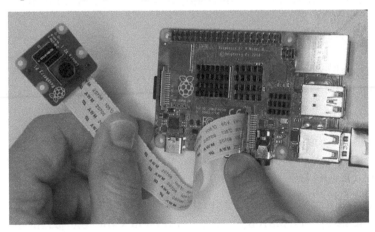

After the camera ribbon clip has been snapped back into place and the seating of the ribbon cable is confirmed stable and in full contact with the camera port connectors, seat the SensorHub back on the GPIO pins. If you haven't used the SensorHub before, refer to Seating the Sensor, on page 63. Be careful not to crease or pinch the Pi Camera's ribbon cable underneath when attaching the SensorHub.

With everything properly seated and installed, power up your Pi.

Camera Activation

Open a Terminal window on your Pi and use the raspi-config program to enable the Pi's camera port that you just attached the camera to:

```
$ sudo raspi-config
```

From the raspi-config screen, select Interface Options, followed by Camera and enable the camera port by selecting the Camera menu option and selecting <Yes> when asked if you would like the camera interface to be enabled.

While you're in the Interface Options screen, double-check to make sure the ARM I2C interface is enabled as well. If you already completed the previous Hue AutoFan or AutoLight projects, this interface should already be enabled. But it can't hurt to double-check, just in case.

With the camera attached and the camera port enabled, you're ready to take your first photo using the Pi Camera.

Take a Picture

With the Terminal window still open, change to your PhotoHook project directory:

```
$ cd /home/pi/projects/photohook
```

Use the raspistill program already installed in the Raspberry Pi OS to take a photo, passing it the file path and name where you want to save the picture:

```
$ raspistill -o test0.jpg
```

After a few seconds, the test0.jpg file should appear in that directory. If raspistill reports an error, verify that your camera ribbon is attached correctly and that the Pi camera port has been enabled.

If you're logged into the Raspberry Pi desktop, navigate to the project directory using the desktop File Manager, and double-click the test0.jpg file to view it in the photo viewer app.

Notice that the default image resolution for photos taken this way is at the full 8-megapixel quality level, 3280 x 2464 pixels. Photos taken at this resolution will be around 4 megabytes in file size.

While that level of image quality is great for still photos that require a high degree of detail, they're rather large to transmit, especially as bot image attachments. Fortunately, raspistill can be configured to reduce both the resolution and quality of the capture. For example, let's take another photo, this time reducing the resolution to 640 x 480 pixels and the image compression quality to 75%. Name this new photo test1.jpg so you can compare it with the higher quality test0.jpg taken earlier:

```
$ raspistill --width 640 --heigh 480 --quality 75 -o test1.jpg
```

Open and examine the test1.jpg image. If it's not still open, reopen the original test0.jpg photo and compare the two images side by side. Unless you zoom in on the photos, you most likely will see little difference between the two in the photo viewer window. However, the difference in file size is huge, with test1.jpg likely coming in at around 203 kilobytes, a difference of over 1,800 percent!

Here's a photo, taken to remotely monitor the results of Chapter 9, Voice IR Control, on page 119, testing at 640x480 resolution and 75% quality. The image looks pretty good even in the format presented in this book.

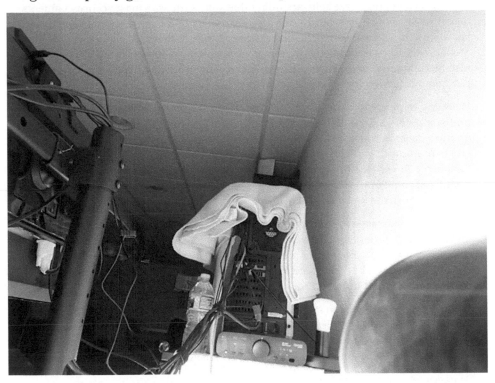

A 200-kilobyte file will transfer significantly faster than a 4,000-kilobyte file and will also consume far less resources and overhead than the higher-resolution image. As such, that is the file quality I recommend using for this project. Of course, if you have a fast Internet connection and an unlimited data plan on your smartphone, you're welcome to use higher resolution for your still-photo captures. Feel free to play with the raspistill values until you attain just the right balance of image quality and file size that best suits your needs.

Motion Detector

For this project, we're going to once again employ the infrared heat sensor on the SensorHub to detect motion, just like we did in Chapter 6, Hue Auto Light, on page 75. Since we already wrote the code for that motion detection routine, we can copy it for this project. The only new routines we need to add are to capture and transmit an image via a Discord webhook.

Joe asks:

Why not use the Motion program?

Motion[a] is a popular, free software utility that works the way most security cameras detect movement. It constantly monitors an incoming image stream for significant variations in pixel values, indicating a state change. When this change is detected, Motion can execute additional programs or scripts in response to the event. It can also be configured to identify motion in specifically defined regions of an image, just like commercial motion-detecting security camera packages.

However, to perform this analysis, Motion requires that the cameras be operational and streaming data at all times, whether there's any movement or not. This means another daemon must be left running to capture and analyze any changes and act on them accordingly. Oh, and the version of Motion that can be installed on the Raspberry Pi via apt install is incompatible with the native Pi camera configuration. Additional configuration overhead is required to make it work, or better yet, you could use an inexpensive webcam and connect it to one of the Pi's USB ports instead.

Since we already have the SensorHub working for our other projects, it will be faster and less computationally expensive to use it in lieu of Motion. However, keep in mind that because the infrared sensor on the Sensorhub requires a heat signature to detect motion, it won't work when operating behind a window or other obstruction blocking the measurement of infrared radiation.

If you decide to give Motion a try via sudo apt install motion, be sure to review the project documentation, adding whatever Python script you want to execute whenever movement is detected to the motion.conf. For example, if you wanted a test.py script to run whenever Motion detected movement, you would add the following line to the motion.conf file:

On_motion_detected python3 test.py

Now, back to our regularly scheduled project.

---

a.   https://motion-project.github.io/index.html

Hook a Photo

If you haven't already done so, create a Discord account and log in to your Discord server. Create a new text channel called #photos using the + icon to the right of the Text Channels listing. Edit the #photos channel settings by selecting the sprocket icon to the right of the #photos channel name. From there, select Integrations and click the Create Webhook button.

Give the new webhook a name like PhotoBot and save the changes.

Select the Copy Webhook URL button and store the generated URL to a safe file. Webhooks are fast and lightweight because they don't require authentication to post messages to a designated channel. That also means anyone with this sensitive webhook URL can do the same thing and spam your channel with unwanted texts.

Now that the basic Discord configuration is done and the webhook URL is activated, we can post a raspistill captured image (or any other file, for that matter) to the #photos text channel.

Post a Picture

Create a new Python script file in the projects/photohook directory using your preferred code editor and name it phototest.py. Add the following lines of Python code, paste your generated webhook URL in place of WEBHOOK_URL_GOES_HERE, and save the file:

```
photohook/phototest.py
import requests

WEBHOOK_URL = 'WEBHOOK_URL_GOES_HERE'
CAPTURE_FILE = r'/home/pi/projects/photohook/capture.jpg'

photo = {'media': open(CAPTURE_FILE, 'rb')}
requests.post(WEBHOOK_URL, files=photo)
```

Open a Terminal window, change to the PhotoHook project directory, and snap a Pi Camera picture using the raspistill app:

```
$ cd ~/projects/photohook
$ raspistill --width 640 --heigh 480 --quality 75 -o capture.jpg
```

Confirm that the new capture.jpg file exists in the directory and run the phototest.py script:

```
$ python3 phototest.py
```

You should see the capture.jpg posted to your Discord #photos channel. Run this script a few more times, capturing different photos with raspistill. Install Discord app on your Android or iOS smartphone, log in, and make sure notifications for the #photos channel are enabled (they should be by default, but check to make sure).

Run the phototest.py and verify that you see the image notification appear on your phone screen.

Verify that everything is working as expected before proceeding to the next and final step in the project.

Movement Notification

We're nearly done. All we need to do now is combine the motionsensortest.py code from Chapter 6, Hue Auto Light, on page 75, with the phototest.py script into a new Python file called photohook.py. Here's what the final contents of that file should contain:

```
photohook/photohook.py
import os
import requests
import smbus
import time

WEBHOOK_URL = 'WEBHOOK_URL_GOES_HERE'
CAPTURE_FILE = r'/home/pi/projects/photohook/capture.jpg'

DEVICE_BUS = 1
DEVICE_ADDR = 0x17

TEMP_REG = 0x01
LIGHT_REG_L = 0x02
LIGHT_REG_H = 0x03
STATUS_REG = 0x04
ON_BOARD_TEMP_REG = 0x05
ON_BOARD_HUMIDITY_REG = 0x06
ON_BOARD_SENSOR_ERROR = 0x07
BMP280_TEMP_REG = 0x08
BMP280_PRESSURE_REG_L = 0x09
BMP280_PRESSURE_REG_M = 0x0A
BMP280_PRESSURE_REG_H = 0x0B
BMP280_STATUS = 0x0C
MOTION_DETECT = 0x0D

bus = smbus.SMBus(DEVICE_BUS)

alert_trigger = False

while True:
    aReceiveBuf = []
    aReceiveBuf.append(0x00)
```

```
    for i in range(TEMP_REG,MOTION_DETECT + 1):
        aReceiveBuf.append(bus.read_byte_data(DEVICE_ADDR, i))
    if aReceiveBuf[MOTION_DETECT] == 1 :
        if alert_trigger != True:
            os.system('raspistill --width 640 --height 480 \
            --quality 75 -o ' + CAPTURE_FILE)
            photo = {'media': open(CAPTURE_FILE, 'rb')}
            requests.post(WEBHOOK_URL, files=photo)
            time.sleep(5) # Wait 5 seconds before capturing another photo
            alert_trigger = True
        else:
            if alert_trigger != False:
                alert_trigger = False

    time.sleep(1)
```

Save and run the script using the usual syntax:

```
$ python3 photohook.py
```

Once the script is running, move in front of the Pi, triggering the infrared sensor. You should see the infrared LED indicator illuminate once it has detected movement. After a few seconds, check your Discord client to confirm that the Pi camera did indeed capture the event and the image was posted to your #photos channel using the PhotoBot's assigned webhook URL.

Test the location and angle of the camera, and position the Sensorhub's orientation so that it optimally captures movement at the angle and location that you want to target.

Set up a Supervisor script if you want PhotoHook to automatically run in the background on your Pi. Refer to the following .conf file:

photohook/photohook.conf
```
[program:photohook]
directory = /home/YOURUSERNAME/projects/photohook
command = /usr/bin/nohup /usr/bin/python3 photohook.py &
user = YOURUSERNAME
environment=HOME="/home/YOURUSERNAME",USER="YOURUSERNAME".
autostart = true
autorestart = true
stdout_logfile = /var/log/supervisor/photohook.log
stderr_logfile = /var/log/supervisor/photohook_err.log
```

Restart the Supervisor service or reboot your Pi, and verify that your PhotoBot is transmitting captured motion photos to your Discord #photos channel.

Next Steps

Congratulations! You just built a motion-detecting security camera with notification updates for a fraction of the cost of similarly featured commercial security cameras. Unlike those third-party black-box alternatives, you also know exactly where and how the images being captured are stored, transmitted, and displayed. Best of all, you don't have to pay a monthly service or subscription to use this custom solution that you built.

If you prefer the simplicity and pop-up notification method of event notification delivery, consider revisiting past projects. Replace the email notification mechanism with a Discord webhook. Or use both of them together for redundant failover messaging of important signals. Combine IR control blasts with a snapshot to provide photographic proof that IR-controlled devices turned on or off as instructed.

Monitor disk space on your Pi and send a text message when storage capacity drops below 10 percent. Merge the PhotoHook code with the Hue AutoLight project to turn on a light in a dimly lit room, pause a few seconds, take a photo, and post to your Discord channel. These projects demonstrate just how much your Pi can do with a small amount of code.

CHAPTER 12

Continuous Improvement

You made it. You. Made. It!

Congratulations on finishing the journey and making the projects in this book. Build upon the knowledge and experience you earned and create your own projects. You have the skills, commitment, and tenacity to implement a range of applications and autonomous hardware and software solutions.

Link together various automation ideas into a cohesive workflow. For example, when I ask my Pi to "Turn on the computer," it incorporates the IR transmitter to turn on my external speakers and the Hue controls to turn on the desk lighting...and turns on my computer, of course.

Once you get more comfortable with the Python scripts used in the book, experiment by changing certain values, modifying routines, and expanding them with other Pip-installed Python libraries. Your only limits are your imagination.

While embarking on your own path, keep the following suggestions in mind.

Maintain

Any well-maintained software project is a constantly evolving one. Stay up-to-date with the latest Python language releases and libraries that you use. Run pip3 install --upgrade on the third-party libraries you installed in these and other Python projects. Such upgrades add features, fix bugs, patch security vulnerabilities, and keep the respective library alive and relevant.

Maintain your own projects, too. As you learn new and improved software-design techniques, apply them to your old code and bring it up to modern standards. Upgrade to newer Pi models as they become available. Faster CPUs, GPUs, network, and more RAM and I/O options are sure to be featured in future iterations of the Pi's product roadmap. They'll also offer more capacity to run larger programs simultaneously and run more complex operations.

Explore

Investigate other Pi-compatible hardware accessories that can bring your ideas to reality. Visit the Raspberry Pi website for a list of vendors who sell add-ons, displays, sensors, actuators, and other electronics that can elevate the Pi far beyond its own designers could imagine. Companies like Adafruit Industries,[1] Chicago Electronic Distributors,[2] and Sparkfun Electronics[3] have online catalogs bursting with Pi-related products, many of which have starter-project articles highlighted on their product pages.

Learn more about the advanced features of Python, such as asynchronous and coroutine programming. Search the web for code snippets that offer clever functions or smarter ways to execute popular routines such as file manipulation and data analysis. Compile a library of these fragments in a notebook. I keep a set of Jupyter[4] notebooks categorized, containing references and clipped Python code samples for quick retrieval and review. The coolest feature of Jupyter notebooks is their ability to execute Python scripts within the notebook itself. Think of a traditional math textbook with the added ability to execute equations and see the results based on different variables in real time, and that is what working with a Jupyter notebook can be like. Microsoft's Visual Studio Code editor has excellent[5] Jupyter notebook support, which is where I call upon my notebook collection most frequently. As for actually finding those code snippet diamonds to store in your own notebook collection, developer-centric websites like DevTalk[6] and StackOverflow[7] can be useful when searching for answers to specific coding questions.

Enjoy

Take the time to enjoy the fruits of your labor. Appreciate how your knowledge and time investment have propelled you into the future, long before these tinkerer projects become commercial products for the masses. Show others what you made and inspire them to build their own projects, just as I hope this book has inspired you to do the same.

Most of all, have fun!

---

1. https://www.adafruit.com/
2. https://chicagodist.com/
3. https://www.sparkfun.com/
4. https://jupyter.org/
5. https://code.visualstudio.com/docs/python/jupyter-support
6. https://www.devtalk.com/
7. https://stackoverflow.com/

Bibliography

[Wal22]   Craig Walls. *Build Talking Apps for Alexa*. The Pragmatic Bookshelf, Raleigh, NC, 2022.

Index

Thank you!

We hope you enjoyed this book and that you're already thinking about what you want to learn next. To help make that decision easier, we're offering you this gift.

Head on over to https://pragprog.com right now, and use the coupon code BUYANOTHER2022 to save 30% on your next ebook. Offer is void where prohibited or restricted. This offer does not apply to any edition of the *The Pragmatic Programmer* ebook.

And if you'd like to share your own expertise with the world, why not propose a writing idea to us? After all, many of our best authors started off as our readers, just like you. With a 50% royalty, world-class editorial services, and a name you trust, there's nothing to lose. Visit https://pragprog.com/become-an-author/ today to learn more and to get started.

We thank you for your continued support, and we hope to hear from you again soon!

The Pragmatic Bookshelf

SAVE 30%!
Use coupon code
BUYANOTHER2022

Raspberry Pi: A Quick-Start Guide (2nd edition)

The Raspberry Pi is one of the most successful open source hardware projects ever. For less than $40, you get a full-blown PC, a multimedia center, and a web server—and this book gives you everything you need to get started. You'll learn the basics, progress to controlling the Pi, and then build your own electronics projects. This new edition is revised and updated with two new chapters on adding digital and analog sensors, and creating videos and a burglar alarm with the Pi camera.

Maik Schmidt
(176 pages) ISBN: 9781937785802. $22
https://pragprog.com/book/msraspi2

Pythonic Programming

Make your good Python code even better by following proven and effective pythonic programming tips. Avoid logical errors that usually go undetected by Python linters and code formatters, such as frequent data look-ups in long lists, improper use of local and global variables, and mishandled user input. Discover rare language features, like rational numbers, set comprehensions, counters, and pickling, that may boost your productivity. Discover how to apply general programming patterns, including caching, in your Python code. Become a better-than-average Python programmer, and develop self-documented, maintainable, easy-to-understand programs that are fast to run and hard to break.

Dmitry Zinoviev
(150 pages) ISBN: 9781680508611. $26.95
https://pragprog.com/book/dzpythonic

Python Brain Teasers

We geeks love puzzles and solving them. The Python programming language is a simple one, but like all other languages it has quirks. This book uses those quirks as teaching opportunities via 30 simple Python programs that challenge your understanding of Python. The teasers will help you avoid mistakes, see gaps in your knowledge, and become better at what you do. Use these teasers to impress your co-workers or just to pass the time in those boring meetings. Teasers are fun!

Miki Tebeka

(116 pages) ISBN: 9781680509007. $18.95

https://pragprog.com/book/d-pybrain

Intuitive Python

Developers power their projects with Python because it emphasizes readability, ease of use, and access to a meticulously maintained set of packages and tools. The language itself continues to improve with every release: writing in Python is full of possibility. But to maintain a successful Python project, you need to know more than just the language. You need tooling and instincts to help you make the most out of what's available to you. Use this book as your guide to help you hone your skills and sculpt a Python project that can stand the test of time.

David Muller

(140 pages) ISBN: 9781680508239. $26.95

https://pragprog.com/book/dmpython

Practical Programming, Third Edition

Classroom-tested by tens of thousands of students, this new edition of the best-selling intro to programming book is for anyone who wants to understand computer science. Learn about design, algorithms, testing, and debugging. Discover the fundamentals of programming with Python 3.6—a language that's used in millions of devices. Write programs to solve real-world problems, and come away with everything you need to produce quality code. This edition has been updated to use the new language features in Python 3.6.

Paul Gries, Jennifer Campbell, Jason Montojo
(410 pages) ISBN: 9781680502688. $49.95
https://pragprog.com/book/gwpy3

Arduino: A Quick-Start Guide, Second Edition

Arduino is an open-source platform that makes DIY electronics projects easier than ever. Gone are the days when you had to learn electronics theory and arcane programming languages before you could even get an LED to blink. Now, with this new edition of the best-selling *Arduino: A Quick-Start Guide*, readers with no electronics experience can create their first gadgets quickly. This book is up-to-date for the latest Arduino boards and for Arduino 1.x, with step-by-step instructions for building a universal remote, a motion-sensing game controller, and many other fun, useful projects.

Maik Schmidt
(322 pages) ISBN: 9781941222249. $34
https://pragprog.com/book/msard2

Exercises for Programmers

When you write software, you need to be at the top of
your game. Great programmers practice to keep their
skills sharp. Get sharp and stay sharp with more than
fifty practice exercises rooted in real-world scenarios.
If you're a new programmer, these challenges will help
you learn what you need to break into the field, and if
you're a seasoned pro, you can use these exercises to
learn that hot new language for your next gig.

Brian P. Hogan
(118 pages) ISBN: 9781680501223. $24
https://pragprog.com/book/bhwb

Learn to Program, Third Edition

It's easier to learn how to program a computer than it
has ever been before. Now everyone can learn to write
programs for themselves—no previous experience is
necessary. Chris Pine takes a thorough, but lightheart-
ed approach that teaches you the fundamentals of
computer programming, with a minimum of fuss or
bother. Whether you are interested in a new hobby or
a new career, this book is your doorway into the world
of programming.

Chris Pine
(230 pages) ISBN: 9781680508178. $45.95
https://pragprog.com/book/ltp3

Kotlin and Android Development featuring Jetpack

Start building native Android apps the modern way in Kotlin with Jetpack's expansive set of tools, libraries, and best practices. Learn how to create efficient, resilient views with Fragments and share data between the views with ViewModels. Use Room to persist valuable data quickly, and avoid NullPointerExceptions and Java's verbose expressions with Kotlin. You can even handle asynchronous web service calls elegantly with Kotlin coroutines. Achieve all of this and much more while building two full-featured apps, following detailed, step-by-step instructions.

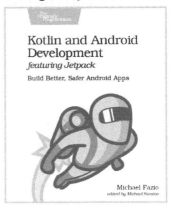

Michael Fazio
(444 pages) ISBN: 9781680508154. $49.95
https://pragprog.com/book/mfjetpack

Modern CSS with Tailwind

Tailwind CSS is an exciting new CSS framework that allows you to design your site by composing simple utility classes to create complex effects. With Tailwind, you can style your text, move your items on the page, design complex page layouts, and adapt your design for devices from a phone to a wide-screen monitor. With this book, you'll learn how to use the Tailwind for its flexibility and its consistency, from the smallest detail of your typography to the entire design of your site.

Noel Rappin
(90 pages) ISBN: 9781680508185. $26.95
https://pragprog.com/book/tailwind

The Pragmatic Bookshelf

The Pragmatic Bookshelf features books written by professional developers for professional developers. The titles continue the well-known Pragmatic Programmer style and continue to garner awards and rave reviews. As development gets more and more difficult, the Pragmatic Programmers will be there with more titles and products to help you stay on top of your game.

Visit Us Online

This Book's Home Page
https://pragprog.com/book/mrpython
Source code from this book, errata, and other resources. Come give us feedback, too!

Keep Up to Date
https://pragprog.com
Join our announcement mailing list (low volume) or follow us on twitter @pragprog for new titles, sales, coupons, hot tips, and more.

New and Noteworthy
https://pragprog.com/news
Check out the latest pragmatic developments, new titles and other offerings.

Save on the ebook

Save on the ebook versions of this title. Owning the paper version of this book entitles you to purchase the electronic versions at a terrific discount.

PDFs are great for carrying around on your laptop—they are hyperlinked, have color, and are fully searchable. Most titles are also available for the iPhone and iPod touch, Amazon Kindle, and other popular e-book readers.

Send a copy of your receipt to support@pragprog.com and we'll provide you with a discount coupon.

Contact Us

Online Orders:	*https://pragprog.com/catalog*
Customer Service:	*support@pragprog.com*
International Rights:	*translations@pragprog.com*
Academic Use:	*academic@pragprog.com*
Write for Us:	*http://write-for-us.pragprog.com*
Or Call:	+1 800-699-7764

Lightning Source UK Ltd.
Milton Keynes UK
UKHW031815040222
398231UK00006B/15